Irrationality and
Transcendence
in Number Theory

Irrationality and Transcendence in Number Theory

David Angell

University of New South Wales, Australia

CRC Press
Taylor & Francis Group
Boca Raton London New York

CRC Press is an imprint of the
Taylor & Francis Group, an **informa** business

A CHAPMAN & HALL BOOK

Cover image: Helena Brusic The Imagination Agency

First edition published 2022
by CRC Press
6000 Broken Sound Parkway NW, Suite 300, Boca Raton, FL 33487-2742

and by CRC Press
2 Park Square, Milton Park, Abingdon, Oxon, OX14 4RN

© 2022 David Angell

CRC Press is an imprint of Taylor & Francis Group, LLC

Library of Congress Cataloging-in-Publication Data

Names: Angell, David (Mathematics), author.
Title: Irrationality and transcendence in number theory / David Angell, University of New South Wales, Australia.
Description: First edition. | Boca Raton : C&H/CRC Press, 2022. | Includes bibliographical references and index.
Identifiers: LCCN 2021037176 (print) | LCCN 2021037177 (ebook) | ISBN 9780367628376 (hardback) | ISBN 9780367628758 (paperback) | ISBN 9781003111207 (ebook)
Subjects: LCSH: Irrational numbers. | Transcendental numbers. | Number theory.
Classification: LCC QA247.5 .A54 2022 (print) | LCC QA247.5 (ebook) | DDC 512.7/3--dc23
LC record available at https://lccn.loc.gov/2021037176
LC ebook record available at https://lccn.loc.gov/2021037177

ISBN: 978-0-367-62837-6 (hbk)
ISBN: 978-0-367-62875-8 (pbk)
ISBN: 978-1-003-11120-7 (ebk)

DOI: 10.1201/9781003111207

Publisher's note: This book has been prepared from camera-ready copy provided by the authors.

In memory of

Dr David K. Crooke

who would have loved to read this book.

Contents

CHAPTER 5 ▪ HERMITE'S METHOD FOR TRANSCENDENCE 109

CHAPTER 6 ▪ AUTOMATA AND TRANSCENDENCE 147

Foreword

H OW CAN WE COMPREHEND the mysteries of the irrational? Irrational
numbers are almost as old as mathematics. The School of Pythagoras in
the 5th century BC believed that "all is number" and rational. It is reported
that the Pythagoreans were at sea when the discovery of incommensurable
ratios was made and that the discoverer was thrown overboard for violating
their code. Geometric construction problems, such as squaring the circle and
trisecting angles, led to more questions about algebraic equations. Going even
further, transcendental numbers beyond the power of algebraic numbers, were
recognised by Euler in the 18th century.

Irrationality and Transcendence in Number Theory deals with the clas-
sification of numbers from the time of Pythagoras to the present day. The
study of irrational and transcendental numbers has been marked by seem-
ingly impossible problems and striking ideas whose ripples spread far beyond
the realm of numbers. I recall the excitement about one recent breakthrough.
In 1966, Alan Baker made a major discovery in transcendental number theory.
He applied it to obtain a new class of transcendental numbers and went on to
develop quantitative versions and solve many classical diophantine equations.
Alan Baker visited Australia and lectured on his work. He was a master of
precision and elegance.

Kurt Mahler was one of the first distinguished mathematicians invited to
start the new research institute at the Australian National University in the
1960s. I was fortunate to hear Mahler expound Alan Baker's work at the
Australian Mathematical Society Summer Research Institute held at the University of Tasmania in 1970. The lectures were called simply, "A theorem of
A. Baker". Mahler was a meticulous lecturer. He started and finished each
lecture with extraordinary punctuality. As he spoke, he wrote a succinct exposition on the blackboard, with the key points neatly placed in order in his
characteristic rectangular boxes. His enthusiasm for mathematics inspired a
generation of number theorists in Australia, led by Alf van der Poorten. I
joined this group when I took up a lectureship at the University of New South
Wales in 1972, and at that time, I was introduced properly to transcendence
and Mahler's method.

Irrationality and Transcendence in Number Theory tells the story from its
origins in the discovery of irrational numbers to the ideas behind the work of
Baker and Mahler on transcendence. The story focuses on important themes

involving irrationality, algebraic and transcendental numbers, continued fractions and Diophantine approximation. These topics make an excellent introduction to modern number theory for advanced undergraduates and early postgraduates.

The book has some unusual and valuable features. Transcending the definitions, theorems and proofs, great care is taken to explain the ideas and illustrate them with examples and well–chosen exercises. This enriches the story, and draws in some material which is not found in more traditional didactic approaches.

One of the themes is Hermite's method. Hermite's work in the nineteenth century introduced so–called auxiliary functions. Hermite used his method to prove that e is transcendental, paving the way for further progress by Lindemann, Gelfond, Schneider and, eventually, Baker. Hermite's method for proving irrationality and transcendence is discussed in two chapters, and the irrationality and transcendence of e and π and much more are demonstrated. The treatment explains some of the mystery behind the proofs and shows how the ideas developed.

Continued fractions and their approximation properties are developed with a little help from Sherlock Holmes. Again, there are some unusual additions. There is a sketch of Apéry's miraculous proof that $\zeta(3)$ is irrational, first announced in 1978. In addition, there is an account of generalised continued fractions, illustrated by Lambert's proof that π is irrational and more.

A final theme revolves around transcendence and computability. This theme is a tribute to Mahler's method in transcendence theory, and the account describes the examples that inspired Mahler to develop his method in the 1930s. Later, Mahler's examples were linked to finite automata, generating a new field of study on the computability of decimal expansions. Taking these ideas further, Adamczewski and Bugeaud [1] proved in 2007 that the decimal expansion of an algebraic irrational cannot be generated by a finite automaton. Despite further remarkable recent progress in 2018 by Adamczewski and Faverjon based on Mahler's method, some simple problems still seem out of reach. For example, how can one approach the widely held belief that the expansions of a number in base 2 and base 10 should have no common structure?

I warmly recommend *Irrationality and Transcendence in Number Theory* as a guide and introduction to number theory. It leads the reader through developments in number theory from ancient to modern times and contains plenty of exercises for practice and problems for exploration. It draws on a wide variety of mathematical techniques, briefly summarised in appendices. The book adopts the tone of a kindly teacher but one not afraid to challenge. It will prepare the reader to appreciate the recent breakthroughs in Baker's method and Mahler's method and tackle the mysteries of the irrational.

John Loxton
13 June 2021

Preface

IRRATIONALITY AND TRANSCENDENCE is a study whose roots go back to about 500 BC, when Pythagoras or one of his followers proved that, contrary to "common sense", certain ratios of geometric quantities cannot be expressed as fractions in whole numbers. While the Ancient Greeks succeeded in proving various surd expressions to be irrational, little further progress was made until the eighteenth century, when Euler and Lambert proved the irrationality of e, π and related numbers. We look first at more modern proofs of these results, deferring Lambert's work until later.

The question of whether a number is algebraic or transcendental – that is, whether it is or is not a root of a polynomial with integer coefficients – is deeper and harder than that of irrationality. After giving a survey of the basic ideas regarding algebraic numbers, we shall prove the existence of transcendentals, firstly (following Cantor) without exhibiting any particular example! The simplest approach to showing that a specific number is transcendental is to study its approximations by rational numbers; continued fractions provide an important tool for doing so. Taking another look at e and π, we shall adapt Hermite's method to prove the transcendence of these numbers.

A (relatively) recent and fascinating topic connects transcendence with deterministic finite automata, a kind of very elementary theoretical computing device. Ideas concerning such automata can be used to investigate the transcendence of numbers that display some sort of "pattern" in their decimal expansions or continued fractions.

The principal aim in writing this book was to present some of the most fundamental techniques for proving numbers irrational or transcendental to students in their later undergraduate or early postgraduate years. These techniques range from the earliest ideas, which may be described as purely numerical (actually, in their original version, geometrical), through to the calculus–based approaches of the nineteenth and twentieth centuries and the mid–to–late twentieth century links with automata theory. I hope that the book will also communicate to readers the delightful, elegant and often surprising nature of many aspects of the subject, and that the expositions given before the formal proofs of some of the harder results will provide a flavour of the process of mathematical discovery, which is invariably more exciting than the mere contemplation of a polished answer.

PREREQUISITES FOR READING THIS BOOK

One of the most fascinating aspects of this subject is that it uses techniques from widely diverse areas of mathematics: number theory, calculus, set theory, complex analysis, linear algebra, order structures, and the theory of computation will all be touched upon. I am firmly convinced that it is not only possible, but appropriate, for readers without any specialist background in these areas to take them on trust in studying irrationality and transcendence. The necessary details have been provided in summary form as appendices to each chapter. For readers who have already studied these topics, the appendices should, in most cases, serve as a reminder, or a checklist, of what prior knowledge is assumed. Those who have not should carefully study the facts listed and, in particular, should ensure that they understand both the assumptions and the conclusion of all theorems given. In the appendices, proofs are provided only in two circumstances: where the specific result needed, though implicit in references cited, may be hard to locate explicitly; and where I could not bring myself to omit an attractive and illuminating line of argument. In some cases, the knowledge set out in the appendices exceeds what the reader requires, and is provided as an encouragement for further study by those readers who so wish.

The content of the book was originally developed as a set of lecture notes for honours level (fourth-year undergraduate) studies in mathematics at the University of New South Wales, Sydney, Australia. It was expected that most students would be largely familiar with the necessary prior knowledge (with the possible exception of the theory of computation) and would only need the appendices as a brief refresher. However, talented and enthusiastic students in their third year, and even some exceptional students in their second year, occasionally took the course, and these were able without excessive difficulty to pick up what they needed from the appendices. For the convenience of the reader, we give some further details.

- **Basic number theory** – divisibility, primes, modular arithmetic. It would be wise for readers to ensure that they are familiar with the material in the appendix to Chapter 1. The Prime Number Theorem is mentioned once in Chapter 3.

- We frequently use the fundamental **calculus** operations of differentiation and integration, notably including integration by parts and estimation of integrals. It is necessary to understand the relative sizes of n^a, b^n and $n!$ for large n. One important proof uses the Mean Value Theorem.

- Elementary **set theory** is really only employed in its capacity as the fundamental language of mathematics. In a couple of places, the idea of a countable set is used: the reader who understands this as a set that can be listed in a finite or infinite sequence, and who knows (or is willing to accept) that the set of complex numbers is not countable, need go no further. For those who would be interested in a somewhat more formal approach, additional details are provided in appendices 3.1 and 4.3.

- **Complex analysis** is used in proving the transcendence of π, section 5.2; however, little is required except the integration of analytic functions, which is effectively identical with the integration of real functions. A small number of results concerning Taylor series and analytic functions are used in Chapter 6.

- A small amount of **linear algebra** and **group theory** is needed in Chapter 3 to prove certain essential results about algebraic numbers and algebraic integers. Exceptionally, an appendix is not provided in this case, as the amount of background required is too great to fit within a brief account. Many readers, however, will have met vector spaces and linear algebra in their studies and should have little difficulty in following our arguments. If necessary, the proofs in section 3.1 may be passed over, as long as the results, which are entirely comprehensible without linear algebra, are read and understood.

- Some basic properties of **order structures** are used in proving Lindemann's Theorem and are explained in appendix 5.3.

- Chapter 6 relates certain ideas in the **theory of computation** to transcendence proofs. This is an area that is very much less likely to form part of an undergraduate mathematics curriculum than those listed above. A couple of formal definitions are given in appendix 6.1 (mainly because I find them interesting); however, for the purposes of the main text, the informal ideas given at the start of section 6.1 are amply sufficient and should be easily absorbed by readers.

- Finally, a few very short appendices contain certain elementary facts about inequalities, solutions of simultaneous equations and other such matters.

Besides this, it is expected that readers will have a basic familiarity with the aims and ideas, and some of the essential techniques of mathematical proof. In particular, proof by contradiction will frequently be important. This is too vast, and too independent, a subject to be combined with the present work, and readers who need to improve their background in this area are cordially recommended to the texts by Polya [52], Franklin and Daoud [26] and Solow [60] which are listed in the bibliography, or to the many other works which have been published recently in this field.

In writing this book, I have striven in many places to simulate the tone of a class discussion. I trust that my expositions of the thinking behind certain difficult arguments, which in some cases almost amount to proving the same result twice, will not be attributed to the woolliness of the author's mind or the flabbiness of his prose, but will help readers to attain an intuitive understanding of the subject which may not always be provided by a text which confines itself to a strict and unvarying alternation of theorem and proof.

For similar reasons, I have not always kept my eye firmly on the principal aims of this book but have availed myself of frequent opportunities to go off on tangents. Some of these asides provide historical background; some are discursions into other areas of mathematics. I make no apology for this; in my view, a higher–level undergraduate course should seek not merely to deepen but also to broaden students' mathematical culture and education.

Each chapter contains exercises to help readers develop their skills further, including many which, I hope, will be found entertaining and thought–provoking. I have not given a great number of "drill" exercises, since readers should be able to provide their own where needed. For instance, the first exercise in Chapter 4 asks for the continued fraction of just one rational number: I feel confident that readers who need more practice in this technique will recognise the fact and will be able to select their own examples.

ACKNOWLEDGMENTS

I would like to acknowledge the three referees who read the first draft of this book before it was accepted for publication. One of them, in particular, very obviously read the manuscript with great care and made many valuable suggestions. While I fully understand and accept the necessity of anonymous refereeing, I regret that I am unable to mention this person by name. I hope that if they read these lines, they will recognise themselves and accept my thanks for their contribution to the final shape of this book.

For introducing me to some of the topics of this book, especially those of Chapter 6, and for his kind Foreword, I am very grateful to John Loxton. My colleagues Igor Shparlinski, Wah Guan Lim and Jim Franklin at UNSW gave help and advice on various points. To Tom Petsinis, mathematician and poet, thanks for allowing me to quote from his work in the latter field.

It was a pleasure to work with Taylor & Francis staff Callum Fraser and Mansi Kabra (editorial) and Shashi Kumar (technical), as well as the production staff who saw the book into print. I had great difficulty in choosing between the many wonderful ideas for the cover image provided by Helena Brusic of The Imagination Agency.

I am indebted to various students who have participated in my honours Irrationality and Transcendence courses over the years for their enthusiasm, their comments on items that were not explained optimally, and their solutions to problems, which often improved upon my own.

And, not least, my love and gratitude to Suzanne Gapps, who never failed in patience while I was writing this book.

David Angell
Sydney, Australia
July 2021

Author

David Angell studied mathematics at Monash University and the University of New South Wales, Australia, earning a PhD from the latter institution with a thesis on Mahler's method in transcendence theory. He has been a member of the academic staff in the School of Mathematics at UNSW since 1989 and has consistently received glowing evaluations of his teaching both from colleagues and students. Dr. Angell has taught a wide variety of mathematics subjects, but his favourites have always been number theory and discrete mathematics. He is particularly interested in teaching students to produce proofs and other mathematical writing which are clearly expressed, logically impeccable and engaging for the reader. He is strongly committed to extension activities for secondary school students. He has, for many years, been the problems editor for *Parabola*, the online mathematics magazine produced by UNSW, as well as a contributor of a number of articles to the magazine. He has also given talks on a wide variety of topics to final–year secondary students.

Beyond mathematics, Dr. Angell is an enthusiast for wilderness activities and has undertaken expeditions in Australia, Greenland, Nepal, Morocco and many other areas. He is a keen amateur musician and is the founding conductor of the Bourbaki Ensemble, a chamber string orchestra based in Sydney, Australia.

CHAPTER 1

Introduction

*It can be of no practical use
to know that π is irrational, but if we can know,
it surely would be intolerable not to know.*

E.C. Titchmarsh

*Die ganzen Zahlen hat der liebe Gott gemacht,
alles andere ist Menschenwerk.*

Leopold Kronecker

T HOUGH THE ORIGINS of number concepts must largely be a matter of speculation, it seems clear that people must from a very early period have seen the necessity of counting collections of objects. It seems clear also that some kind of numerical abstraction of the counting numbers (the concept of "five", rather than "five fingers", "five sheep" or "five days") must have been understood a long time ago. Probably the next development of numerical technique would have been connected with division of a quantity into equal parts and would have led to the use of fractions. (Zero and negative numbers did not appear until many centuries later.) Early Greek geometers seem to have assumed that any two line segments are *commensurable*: that is, that a "common measure" can always be found such that each of two given segments is an integral multiple of the common measure. This is, in effect, to assume that the ratio of lengths of line segments is always a rational number. In the time of Pythagoras (*ca.* 570–490 BC) it was discovered that this assumption is invalid in the case where the segments in question are the side and diagonal of a square. This discovery had various consequences: one, the development of an improved theory of proportion (Eudoxus, *ca.* 408–355 BC) which applied equally to incommensurable and to commensurable lines; another, the study of irrational numbers.

Assuming that the positive integers $1, 2, 3, \ldots$ are known we can construct the (signed) integers, the rationals, the reals and the complex numbers. Each construction may be based on the previously known numbers, and at every stage we should prove that the necessary algebraic laws hold (including some

DOI: 10.1201/9781003111207-1

that we already know to hold in the earlier cases). Indeed, we can go back even further than this and define the positive integers in terms of set theory, here also taking care to prove the essential properties of the integers. For the present topic, however, this development is of no importance and we shall work the other way around: we shall assume that we know everything(!) about the complex numbers $\mathbb{C}$, and shall then define rational and irrational numbers as subsets of $\mathbb{C}$.

Definition 1.1. *A* **rational number** *is a number of the form p/q, where p and q are integers and q is not zero. An* **irrational number** *is any (complex) number which is not rational.*

It is well known that any rational number can be written *uniquely* in the form p/q, where p and q are integers with no common factor and q is positive. We shall frequently assume when we refer to a rational number p/q that these properties already hold. Any non–real complex number is necessarily irrational; for this reason we shall initially concern ourselves mainly with real numbers. However, it turns out that some problems involving real numbers can be significantly simplified by writing them in terms of complex numbers: an example will be seen in exercise 1.12 at the end of this chapter. Moreover, in Chapter 3 we shall need to consider complex numbers in connection with a further subdivision of the irrational numbers.

1.1   IRRATIONAL SURDS

The following result is well known, and was, essentially, proved by Pythagoras or one of his followers.

Theorem 1.1. $\sqrt{2}$ *is irrational.*

Proof by contradiction. Suppose that $\sqrt{2} = p/q$, where p and q are integers with no common factor, and with $q \neq 0$. Squaring both sides and multiplying by q^2, we have $p^2 = 2q^2$. Thus p^2 is even and so p is even, say $p = 2r$. Substituting for p gives $q^2 = 2r^2$ and so q is even. Thus p and q have a common factor of 2, and this contradicts our initial assumption. Therefore, $\sqrt{2}$ is irrational.

Plato records that his teacher Theodorus proved the irrationality of $\sqrt{n}$ for n up to 17. Historians of mathematics have wondered why he stopped just here; the question is made harder by the fact that we don't know exactly how Theodorus' proof ran. The following proof of the irrationality of $\sqrt{n}$ for certain values of n suggests a possible reason for stopping just before $n = 17$.

First, if $n = 4k$, then the irrationality of $\sqrt{n}$ is equivalent to that of $\sqrt{k}$; and if $n = 4k + 2$, then the method used above for $n = 2$ can be employed with only minor changes. So we concentrate on odd values of n. If n is odd and $\sqrt{n} = p/q$, then $nq^2 = p^2$ and p and q must both be odd; substituting $p = 2r + 1$ and $q = 2s + 1$ and rearranging yields

$$4n(s^2 + s) - 4(r^2 + r) + n - 1 = 0 .$$

Consider the case $n = 4k + 3$. Cancelling 2 from the above equation gives

$$2n(s^2 + s) - 2(r^2 + r) + 2k + 1 = 0 ,$$

which is clearly impossible as the left–hand side is odd. This method does not work directly for $n = 4k + 1$, so we consider as a subsidiary case $n = 8k + 5$. Substituting as above and cancelling 4 we obtain

$$n(s^2 + s) - (r^2 + r) + 2k + 1 = 0 ;$$

but as $r^2 + r$ and $s^2 + s$ are both even, this is again impossible.

The remaining possibility is that $n = 8k + 1$; but it appears that this case has to be split up into still further subcases, and the proof becomes much more complicated (try it!), so we shall stop here. Therefore, we have proved the following.

Theorem 1.2. *If n is not of the form $4k$ or $8k + 1$, then $\sqrt{n}$ is irrational. If $n = 4k$, then $\sqrt{n}$ is irrational if and only if $\sqrt{k}$ is irrational.*

How far can we get using this result? We have settled every case except for $n = 4^m(8k + 1) = 1, 4, 9, 16, 17, 25, 33, \ldots$. Of these, the first four obviously have rational square roots, and so the smallest undecided case is $\sqrt{17}$. Thus, if the above was Theodorus' method of proof, it would have been quite reasonable for him to stop before reaching $n = 17$. Hardy and Wright [29], section 4.5, comment further that this proof "[depends] essentially on the distinction between odd and even, a matter of great importance in Greek mathematics".

Comment. The working in the above proof can be simplified somewhat by using modular arithmetic; but this concept was not available to the ancient Greeks, being introduced into number theory by Gauss, some 24 centuries after the time of Pythagoras.

Our first proof of the irrationality of $\sqrt{2}$ can be slightly reorganised in a way which admits an important generalisation.

Lemma 1.3. The Rational Roots Lemma. *If a rational number p/q, where p and q have no common factors, is a root of the polynomial equation*

$$a_n z^n + a_{n-1} z^{n-1} + \cdots + a_1 z + a_0 = 0$$

with integer coefficients $a_0, a_1, \ldots, a_{n-1}, a_n$, then p is a factor of a_0 and q is a factor of a_n.

Proof. Substituting $z = p/q$, multiplying by q^n and rearranging, we have

$$a_n p^n = -a_{n-1}p^{n-1}q - \cdots - a_1 pq^{n-1} - a_0 q^n$$
$$= -q(a_{n-1}p^{n-1} + \cdots + a_1 pq^{n-2} + a_0 q^{n-1}) .$$

Hence $q \mid a_n p^n$; but q and p have no common factor, and so $q \mid a_n$. Similarly $p \mid a_0$.

Corollary 1.4. *Another proof of the irrationality of $\sqrt{2}$*. Suppose, on the contrary, that $\sqrt{2} = p/q$. Noting that $\sqrt{2}$ is a root of $z^2 - 2 = 0$ and applying the above lemma, we have $q = 1$ and so $\sqrt{2}$ is an integer. But since $1 < \sqrt{2} < 2$ this is clearly not true.

Examples.

- Let $f(z) = 3z^3 + 4z^2 + 5z + 6$. Suppose that p/q is a rational root of f, with p and q coprime. Then $p \mid 6$ and $q \mid 3$; without loss of generality q is positive, so the possibilities are

$$\frac{p}{q} = 1, 2, 3, 6, -1, -2, -3, -6, \frac{1}{3}, \frac{2}{3}, -\frac{1}{3}, -\frac{2}{3} .$$

  It is clear that f has no positive roots, and a bit of calculation also eliminates the negative numbers from the above list. So f has no rational roots.

- Let $f(z) = 168z^3 - 133z + 275$. If we use the above approach we will have 192 potential roots to check! (*exercise:* confirm this). However, if we apply the method rather than the result of the Rational Roots Lemma, a little ingenuity will give a rapid solution. Suppose that f has a root p/q, where p and q are relatively prime integers. Then

$$168p^3 - 133pq^2 + 275q^3 = 0 .$$

We have

  - ∘ $7 \mid 168$ and $7 \mid 133$ and $7 \nmid 275$, so $7 \mid q$;
  - ∘ $q^2 \mid 168p^3$ and p^3, q^2 have no common factor, so $q^2 \mid 168$;

  therefore $49 \mid 168$, which is not true. Hence f has no rational roots.

Definition 1.2. *A polynomial is said to be* **monic** *if its leading coefficient (the coefficient in the term of highest degree) is 1.*

Theorem 1.5. Roots of monic polynomials. *Let α be a root of a monic polynomial with integer coefficients. Then α is either integral or irrational.*

Proof. Suppose that α satisfies

$$a_n z^n + a_{n-1}z^{n-1} + \cdots + a_1 z + a_0 = 0$$

with $a_n = 1$ and that $\alpha = p/q$. By the lemma we have $q \mid a_n$, so $q = 1$ and α is an integer.

Definition 1.3. *Any root of a monic polynomial with integer coefficients is called an* **algebraic integer**.

Notes.

- Any "ordinary" integer $n \in \mathbb{Z}$ is an algebraic integer, as it is a root of the monic polynomial equation $z - n = 0$.

- If we want to emphasize the difference between algebraic integers and "ordinary" integers, we shall refer to the latter as **rational integers**. The term is justified by the following result.

  Theorem 1.6. *A complex number is a rational integer if and only if it is both rational and an (algebraic) integer.*

  Proof. The forward implication is clear; the converse is just a restatement of the previous theorem.

- If α is a complex number such that

$$a_n\alpha^n + a_{n-1}\alpha^{n-1} + \cdots + a_1\alpha + a_0 = 0$$

  for some rational integers $a_0, a_1, \ldots, a_{n-1}, a_n$, not all zero, (such a number is called **algebraic**), then $a_n\alpha$ is an algebraic integer.

- If α and β are algebraic integers, then so are $\alpha + \beta$, $\alpha - \beta$ and $\alpha\beta$.

- If $\alpha_0, \alpha_1, \ldots, \alpha_{n-1}$ are algebraic integers and β is a root of

$$z^n + \alpha_{n-1}z^{n-1} + \cdots + \alpha_1 z + \alpha_0 \,,$$

  then β is an algebraic integer. In particular, if n is a positive rational integer and α is an algebraic integer, then any (possibly complex) value of $\sqrt[n]{\alpha}$ is an algebraic integer.

These results (which we shall prove later) make it easy to show that various simple expressions consisting of radicals are irrational. For example, $\sqrt{29} - \sqrt{23}$ is the difference of two algebraic integers, and therefore is either integral or irrational; but

$$0 < \sqrt{29} - \sqrt{23} = \frac{6}{\sqrt{29} + \sqrt{23}} < \frac{6}{5 + 4} < 1 \,,$$

so $\sqrt{29} - \sqrt{23}$ cannot be an integer and must be irrational. For a slightly harder example, consider the polynomial

$$p(z) = z^3 - \sqrt{2}\, z^2 + \sqrt[3]{4}\, z - \sqrt[5]{6} \,.$$

The derivative $p'(z) = 3z^2 - 2\sqrt{2}\, z + \sqrt[3]{4}$ is a quadratic with discriminant

$$\Delta = b^2 - 4ac = 8 - 12\sqrt[3]{4} \,,$$

which is negative; so $p'(z)$ is always positive, $p(z)$ is always increasing, and $p(z)$ has just one real root α. It is easy to see that the coefficients of p are algebraic integers, and so by the last property on the previous page α is also an algebraic integer. A little thought shows that

$$p(1) = 1 - \sqrt{2} + \sqrt[3]{4} - \sqrt[5]{6} < 1 - \tfrac{4}{3} + \tfrac{5}{3} - \tfrac{4}{3} = 0 \ ,$$

$$p(2) = 8 - 4\sqrt{2} + 2\sqrt[3]{4} - \sqrt[5]{6} > 8 - 8 + 2 - 2 = 0$$

and so $1 < \alpha < 2$. Since α is an algebraic integer but not a rational integer, it is irrational.

1.2   IRRATIONAL DECIMALS

The following well–known result characterises rational numbers in terms of their decimals. Note that the eventually periodic decimal expansions include the finite expansions, for instance, $0.123 = 0.123000 \cdots = 0.122999 \cdots$.

Theorem 1.7. Rationality of decimals. *A real number α is rational if and only if it has an eventually periodic decimal expansion.*

Proof. Firstly, suppose that α has an eventually periodic expansion. Without loss of generality we may assume that $0 < \alpha < 1$, say

$$\alpha = 0.a_1 a_2 \cdots a_s b_1 b_2 \cdots b_t b_1 b_2 \cdots b_t b_1 b_2 \cdots \ .$$

Let a and b be the non–negative integers with digits $a_1 a_2 \cdots a_s$ and $b_1 b_2 \cdots b_t$ respectively; then

$$\alpha = \frac{a}{10^s} + \frac{b}{10^{s+t}} + \frac{b}{10^{s+2t}} + \cdots = \frac{a}{10^s} + \frac{b}{10^{s+t}} \frac{1}{1 - 10^{-t}} \ ,$$

which is rational. Conversely, suppose that $\alpha = p/q$ is rational, and initially assume that neither 2 nor 5 is a factor of q. Choose $t = \phi(q)$, where ϕ is Euler's function: see definition 1.6 in the appendix to this chapter. By Euler's Theorem we have

$$10^t \equiv 1 \pmod{q}$$

and so q is a factor of $10^t - 1$, say $10^t - 1 = qr$. Hence we can write

$$\alpha = \frac{pr}{10^t - 1} = a + \frac{b}{10^t - 1} \ ;$$

here we have used the division algorithm to guarantee that $0 \le b < 10^t - 1$. We can thus write b as a number of t digits, say $b = b_1 b_2 \cdots b_t$; it is possible that b_1 is zero. Similarly, write $a = a_1 a_2 \cdots a_s$. Then

$$\alpha = a + \frac{b}{10^t} + \frac{b}{10^{2t}} + \cdots = a_1 a_2 \cdots a_s . b_1 b_2 \cdots b_t b_1 b_2 \cdots b_t b_1 b_2 \cdots \ ,$$

and we see that α has an eventually periodic decimal expansion. To complete the proof we must also consider the case when q has 2 or 5 as a factor. Let $q = 2^m 5^n q'$, where neither 2 nor 5 is a factor of q'; then

$$10^{m+n}\alpha = \frac{2^n 5^m p}{q'} = \frac{p'}{q'} ,$$

say; by the previous argument, the decimal expansion of $10^{m+n}\alpha$ is eventually periodic. The expansion of α contains exactly the same digits (with the decimal point shifted $m + n$ places), so it too is eventually periodic.

Alternative proof (sketch). To show that every rational number $\alpha = p/q$ has an eventually periodic decimal expansion, suppose without loss of generality that $0 \leq \alpha < 1$, and consider how to compute the expansion

$$\alpha = 0.a_1 a_2 a_3 \cdots$$

by division. We divide $10p$ by q; the quotient is a_1 and the remainder, say, p_1. Dividing $10p_1$ by q gives quotient a_2 and remainder p_2, and so on. Since division by q gives only a finite number of possible remainders, the remainder p_k must at some stage be the same as a previous remainder p_j. From this point on the whole procedure repeats and we have $a_k = a_j$, $p_{k+1} = p_{j+1}$, $a_{k+1} = a_{j+1}$ and so on. **Exercise.** Write out this proof in more detail.

Examples.

- Consider the number

$$\alpha = \sum_{k=0}^{\infty} 10^{-k^2} = 1.1001000010000001000000000100000 \cdots .$$

If the decimal expansion is eventually periodic with period length t, then the periodic part must contain t consecutive zeros, and therefore must be entirely zero – which is obviously false. So the decimal is not eventually periodic, and α is irrational. Similar examples (of which we shall later learn a good deal more) are

$$\sum_{k=0}^{\infty} 10^{-2^k} = 0.110100010000000100000000000000010000 \cdots$$

and **Liouville's number,**

$$\sum_{k=1}^{\infty} 10^{-k!} = 0.110001000000000000000010000 \cdots .$$

- The **Champernowne constant** is obtained by stringing together all the positive integers in their natural order:

$$\xi = 0.12345678910111213141516 \cdots .$$

This number is irrational by the same argument as above.

- The **Thue sequence** is defined inductively as follows: $a_0 = 0$, and

$$a_{2k} = a_k , \quad a_{2k+1} = 1 - a_k = \begin{cases} 1 & \text{if } a_k = 0 \\ 0 & \text{if } a_k = 1 \end{cases}$$

for any $k \geq 0$. (Note that the first equation is consistent when $k = 0$.) Taking this sequence as the digits of an infinite decimal, we obtain the number

$$\tau = 0.11010011001011010010110\cdots .$$

This is not an eventually periodic expansion. **Proof**. Suppose the contrary. Then there exist integers $t \geq 1$ and $N \geq 0$ such that $a_{n+t} = a_n$ for all $n \geq N$; also, we may assume that t is the smallest positive integer for which such an N exists. First, observe from the definition that if m is even, then $a_{m+1} = 1 - a_m$. Now let n be an integer not less than N, and such that $n + t$ is even. Then by using the assumed periodicity of the Thue sequence, the definition and the previous observation, we have

$$a_{2n+t+1} = a_{2n+2t+1} = 1 - a_{n+t} = a_{n+t+1} = a_{2n+2t+2} = a_{2n+t+2} ,$$

and so $2n + t + 1$ cannot be even. Thus t is even, say $t = 2s$, and for all $n \geq N$, we have

$$a_{n+s} = a_{2n+2s} = a_{2n+t} = a_{2n} = a_n .$$

But this is impossible as t is the *least* integer with such a property, while clearly $s < t$. The contradiction shows that the sequence is not eventually periodic.

Corollary 1.8. *The number τ is irrational.*

Comment. An alternative characterisation of the Thue sequence: a_k is the parity of the binary representation of the integer k. The sequence is also known as the Prouhet–Thue–Morse sequence; it has connections with a wide variety of fields including harmonic analysis, dynamical systems, differential geometry... and, potentially, penalty shoot–outs in football [49].

1.3   IRRATIONALITY OF THE EXPONENTIAL CONSTANT

Once we get beyond radical expressions and decimals, irrationality proofs, for the most part, become significantly harder. A notable exception is the irrationality of the exponential constant e. Apart from the intrinsic interest of the result, its proof provides our first glimpse of an idea which will recur again and again in irrationality arguments, and which we shall employ extensively in Chapters 2 and 5.

Theorem 1.9. *The exponential constant e is irrational.*

Proof. Assume that $e = p/q$ is rational. That is,

$$\frac{p}{q} = 1 + \frac{1}{1!} + \frac{1}{2!} + \frac{1}{3!} + \cdots ,$$

and for any positive integer n, we have

$$\frac{p\,n!}{q} = n! + \frac{n!}{1!} + \frac{n!}{2!} + \cdots + 1 + R ,$$

where R (which depends on n) is given by

$$R = \frac{n!}{(n+1)!} + \frac{n!}{(n+2)!} + \cdots .$$

We can estimate R in terms of a geometric series:

$$R = \frac{1}{n+1} + \frac{1}{(n+1)(n+2)} + \cdots < \frac{1}{n+1} + \frac{1}{(n+1)^2} + \cdots = \frac{1}{n} . \quad (1.1)$$

In particular, choose $n = q$. Then

$$R = \frac{p\,n!}{q} - \left(n! + \frac{n!}{1!} + \frac{n!}{2!} + \cdots + 1 \right)$$

is clearly an integer; but using (1.1), we have $0 < R < 1$. This is impossible, and so e is irrational.

Observe that this proof relies essentially on an infinite series for e, and therefore has to involve concepts of calculus. In some sense this may be surprising, as number theory is usually thought of as studying discrete systems while calculus is the science of the continuous; in another sense there should be no surprise, as it is not even possible to define the number e without recourse to calculus techniques. Whether it is in fact a surprise or not, we shall find that many of our future proofs will be expressed in terms of calculus.

1.4 OTHER RESULTS, AND SOME OPEN QUESTIONS

It is known that π is irrational: we shall prove this in the next chapter. It is not hard to see that at least one of the numbers $\pi + e$ and πe must be irrational (in fact, at least one must be *transcendental* – see Chapter 3); although, most likely, both are irrational, this has not been proved for either one individually. As a consequence of a difficult result due to Gelfond and Schneider (Theorem 5.18) we know that e^π is irrational; however it is still unknown whether or not π^e is irrational. It can also be shown that various numbers such as, for example, $e^{\sqrt{2}}$ and $2^{\sqrt{2}}$ are irrational. However, the irrationality of $\pi^{\sqrt{2}}$ and 2^e, and that of the Euler–Mascheroni constant

$$\gamma = \lim_{n \to \infty} \left(1 + \frac{1}{2} + \frac{1}{3} + \cdots + \frac{1}{n} - \log n \right) = 0.57721 \cdots$$

remain undecided. Another problem which has attracted much attention is to investigate the irrationality of the numbers $\zeta(n)$. Here $n \geq 2$ is an integer and ζ is the *Riemann zeta function* defined by

$$\zeta(s) = \sum_{k=1}^{\infty} \frac{1}{k^s} = 1 + \frac{1}{2^s} + \frac{1}{3^s} + \frac{1}{4^s} + \cdots$$

for $s > 1$. By methods of complex integration we can show that if n is even then $\zeta(n)$ is a rational number times π^n, and this is known to be irrational. On the other hand, it is much harder to find out anything of interest about $\zeta(n)$ for odd n. In 1978 the French mathematician R. Apéry sensationally proved that $\zeta(3)$ is irrational. His complicated argument had the appearance of being completely unmotivated, and all of the techniques he had used would have been available two centuries earlier: for these reasons, few people believed that the proof could possibly be correct. Nevertheless it was found possible eventually to confirm all of Apéry's assertions and thereby establish what has been called "a proof that Euler missed". A brief (but not easy!) account of Apéry's work is given in [66].

EXERCISES

1.1 Assuming that $\sqrt{2} = p/q$, simplify $(2q - p)/(p - q)$. Use the result to give an alternative proof of the irrationality of $\sqrt{2}$.

1.2 Generalising the previous question, suppose that $k^2 < n < (k+1)^2$. Show that $\sqrt{n}$ is irrational by assuming that $\sqrt{n} = p/q$ and simplifying the expression $(nq - kp)/(p - kq)$.

1.3 We can show that $\alpha = \sqrt[m]{n}$ is integral or irrational without relying in any way on properties of primes. Suppose that α is rational, $\alpha = p/q$ with q minimal, and write $\beta_k = q^{m-k-1}\alpha^{m-k}$; then prove by induction on k that β_k is an integer for $k = 0, 1, \ldots, m - 1$.

1.4 Let a and b be unequal positive rational numbers. Show that

(a) if $\sqrt{a} - \sqrt{b}$ is rational, then $\sqrt{a}$ and $\sqrt{b}$ are rational;

(b) if $\sqrt[3]{a} - \sqrt[3]{b}$ is rational then $\sqrt[3]{a}$ and $\sqrt[3]{b}$ are rational.

Challenge problem. Show that if a, b are positive rationals, $a \neq b$ and $\sqrt[n]{a} - \sqrt[n]{b}$ is rational, then $\sqrt[n]{a}$ and $\sqrt[n]{b}$ are both rational. It may not be possible to do this with the methods we have introduced so far, though there is a reasonably simple solution using ideas from Chapter 3.

1.5 Find all positive rational r such that $r^{1/r}$ is rational.

1.6 Prove that 1, $\sqrt[3]{2}$, $\sqrt[3]{4}$ are linearly independent over $\mathbb{Q}$. That is, show that the equation $a + b\sqrt[3]{2} + c\sqrt[3]{4} = 0$ has no rational solutions a, b, c other than $a = b = c = 0$.

1.7 Let p/q be a rational number between 0 and 1, with p, q having no common factor. Prove that

$$\left| \frac{1}{\sqrt{2}} - \frac{p}{q} \right| > \frac{1}{4q^2} \, .$$

1.8 Show that there are no positive integers a, b, except for $a = 3$, $b = 2$, such that

$$x = \frac{\sqrt{2} + \sqrt{a}}{\sqrt{3} + \sqrt{b}}$$

is rational.

1.9 Complete the statement and prove it: "if n is a positive integer and m is prime, then $\log_m n$ is irrational unless. . . ".

1.10 Show that if α is a root of the polynomial $az^n + bz + c$, where a, b and c are odd integers and $n \geq 2$, then α is irrational. Generalise.

1.11 Show that $\alpha = \sin \frac{1}{18}\pi$ is irrational.

1.12 Let r be a rational number.

 (a) Show that $e^{r\pi i}$ is an algebraic integer.

 (b) Show that $\cos r\pi$ is either 0, $\pm\frac{1}{2}$, ± 1 or irrational.

 (c) What are the possible rational values of $\cos^2 r\pi$?

1.13 Show that the decimal obtained by concatenating the digits of powers of 2, that is,
$$0.1248163264128256 \cdots$$
is irrational.

1.14 Repeat the previous question for powers of 13. That is, prove that

$$0.1 \, 13 \, 169 \, 2197 \, 28561 \, 371293 \, 4826809 \cdots$$

is irrational.

1.15 Prove that the *Copeland–Erdős constant*

$$\alpha = 0.235711131719232931374143 47 \cdots ,$$

whose decimal is obtained by concatenating the digits of the primes in increasing order, is irrational. You may assume **Bertrand's postulate**:

for each integer $n \geq 2$ there is a prime p such that $n \leq p < 2n$. Although it is (still) traditionally known as a "postulate", this result was proved by Chebyshev in 1850 and is definitely true! See, for example, [2], Chapter 2.

1.16 Let $a_1, a_2, a_3, \ldots$ be a strictly increasing sequence of positive integers, and let α be the real number obtained by concatenating their digits after a decimal point. Prove that if α is rational then there exist constants $c > 0$ and $x > 1$ such that $a_k > cx^k$ for all k.

1.17 If p_k is the kth prime, it can be shown that $p_k / k \log k \to 1$ as $k \to \infty$. (This is related to the Prime Number Theorem: see appendix 3.3.)

Use this and the result of the previous problem to give a proof (different from that in exercise 1.15) that $0.23571113171923293137\cdots$ is irrational.

1.18 Let

$$\alpha = 0.12345678912345678902345678901345678901245\cdots \ ,$$

where the kth digit is the sum of the digits of k, reduced modulo 10.

(a) Show that the decimal never has the same digit more than twice consecutively (and so our basic argument for the irrationality of a decimal, as in the examples on page 7, will not work).

(b) Prove that α is irrational.

1.19 Prove that a real number α is rational if and only if $q! \, \alpha$ is an integer for all sufficiently large integers q. Deduce that $\cos 1$ is irrational.

1.20 By considering the equation $ae + ce^{-1} = b$, show that e is not a *quadratic irrational*; that is, e is not the root of a quadratic polynomial with integer coefficients. Deduce that e^2 is irrational.

1.21 *Generalised "decimal" expansions.* Let $\{\, g_n \,\}_{n=1}^{\infty}$ be a sequence of integers greater than 1, and for each n let a_n be an integer in the range $0 \leq a_n \leq g_n - 1$. Write

$$\alpha = \frac{a_1}{g_1} + \frac{a_2}{g_1 g_2} + \frac{a_3}{g_1 g_2 g_3} + \cdots \ .$$

Suppose that there are infinitely many n such that $a_n \neq 0$, and infinitely many n such that $a_n \neq g_n - 1$; and that for each prime p, infinitely many g_n are multiples of p. Show that α is irrational.

Comment. "Normal" decimal expansions correspond to the case with $g_1 = g_2 = \cdots = 10$, which certainly does not satisfy the condition that infinitely many g_n are multiples of any prime. Thus, this exercise provides a nice complement to Theorem 1.7.

1.22 Show that $\alpha = e^{\sqrt{2}} + e^{-\sqrt{2}}$ is irrational; deduce that $e^{\sqrt{2}}$ is irrational.

1.23 Let P be a polyhedron with *dihedral angles* (that is, angles between adjacent faces) $\alpha_1, \ldots, \alpha_s$ and Q a polyhedron with dihedral angles $\beta_1, \ldots, \beta_t$. It can be shown (see, for example, [2], Chapter 10) that if P can be dissected into finitely many pieces which can be reassembled into (a congruent copy of) Q, then

$$p_1\alpha_1 + \cdots + p_s\alpha_s = q_1\beta_1 + \cdots + q_t\beta_t + r\pi$$

for some integer r and some (strictly) positive integers $p_1, \ldots, p_s$, $q_1, \ldots, q_t$. Use this result to prove that it is impossible to dissect a regular tetrahedron and reassemble the pieces to form a cube.

1.24 Book X of Euclid's *Elements* contains definitions and propositions about irrationality. It is confusing and frustrating to read, partly because there is no algebraic notation, but mainly because in Euclid's terminology (as usually translated into English), a "rational line" is one whose *square* is rational–in–the–modern–sense. Two lines are said to be "commensurable" if their ratio is rational–in–the–modern–sense, and "commensurable in square" if the ratio of their squares is rational–in–the–modern–sense. "Commensurable in square only" means commensurable in square but not commensurable. Bearing all this in mind, express the following, taken from [24], in modern notation, and prove it.

Book X, proposition 73. If from a rational straight line there is subtracted a rational straight line commensurable with the whole in square only, then the remainder is irrational.

1.25 *For readers who are familiar with complex integration and the method of residues.* Use the following steps to show that if n is even then $\zeta(n)$ is a rational multiple of π^n. Also, indicate where the proof breaks down if n is odd.

Let $f(z) = z^{-n} \cot z$, and let C be the square contour in the complex plane having vertical sides at $x = \pm(N + \frac{1}{2})\pi$ and horizontal sides at $y = \pm(N + \frac{1}{2})\pi$.

(a) Show that f has simple poles (that is, poles of order 1) at the points $z = k\pi$ with $k = \pm 1, \pm 2, \ldots$, and find the residue at each of these poles.

(b) Explain why f has a Laurent series in powers of z given by

$$f(z) = b_{n+1}z^{-(n+1)} + \cdots + b_1 z^{-1} + a_0 + a_1 z + \cdots .$$

Use the identity $z^{n+1}f(z)\sin z = z\cos z$ to show that every coefficient b_j is rational. Deduce that

$$\int_C z^{-n} \cot z \, dz = 2\pi i \left(r_n + \frac{2}{\pi^n} \sum_{k=1}^{N} \frac{1}{k^n} \right),$$

where r_n is a rational number.

(c) Use the identity

$$|\cot z|^2 = \frac{\cos^2 x + \sinh^2 y}{\sin^2 x + \sinh^2 y}$$

to show that $|\cot z| \leq 1$ on the vertical sides of the square, and $|\cot z| \leq \coth(\pi/2)$ on the horizontal sides. Deduce that

$$\int_C z^{-n} \cot z \, dz \to 0 \quad \text{as } N \to \infty$$

and hence, finally, that $\zeta(n)$ is a rational multiple of π^n.

APPENDIX: SOME ELEMENTARY NUMBER THEORY

This section contains some basic number–theoretic definitions and results which you ought to know. Proofs in this section are abbreviated or omitted, and you should be able to supply proofs for yourself. If necessary, this material can be found in any work on elementary number theory. The most popular of the classic texts are regularly revised, thereby offering a proven exposition together with additions which bring the content and presentation up to date. From a very crowded field we mention Hardy and Wright [28], [29], Niven and Zuckerman [45], [46] and Baker [10].

Lemma 1.10. The division algorithm. *If a and b are integers with $b > 0$, then there exist integers q and r such that $a = bq + r$ and $0 \leq r < b$.*

Using the division algorithm recursively gives the *Euclidean algorithm* for computing the greatest common divisor of two integers, not both zero.

Lemma 1.11. The Bézout property. *If a and b are integers, not both zero, and g is the greatest common divisor of a and b, then there exist integers x and y such that $ax + by = g$.*

Given specific a and b, you should know how to use the Euclidean algorithm to find g, x and y.

Lemma 1.12. *If a and m have no common factor and $a \mid mn$, then $a \mid n$.*

Definition 1.4. *Let m be a positive integer. We say that integers a and b are* **congruent modulo** m, *written $a \equiv b \pmod{m}$, if $m \mid a - b$.*

To "reduce an integer a modulo m" means to find an integer b such that $a \equiv b \pmod{m}$ and b lies in a "suitable" range, usually $0 \leq b < m$. That this can always be done is a consequence of the division algorithm. Although congruence notation is just another way of expressing a divisibility relation, and in that sense "nothing new", it is very useful because congruence shares many of the basic properties of equality.

Lemma 1.13. Properties of congruence. *Let m be a positive integer.*

- Equivalence properties. *For any integers a, b, c, we have*

    ○ $a \equiv a \pmod{m}$;

    ○ *if* $a \equiv b \pmod{m}$, *then* $b \equiv a \pmod{m}$;

    ○ *if* $a \equiv b \pmod{m}$ *and* $b \equiv c \pmod{m}$, *then* $a \equiv c \pmod{m}$.

- Congruence properties. *If* $a \equiv b \pmod{m}$ *and* $c \equiv d \pmod{m}$, *then*

    ○ $a + c \equiv b + d \pmod{m}$;

    ○ $a - c \equiv b - d \pmod{m}$;

    ○ $ac \equiv bd \pmod{m}$.

    If, also, n is a non–negative integer, then

    ○ $a^n \equiv b^n \pmod{m}$.

- A cancellation property. *Let a, b, m and s be integers. If $sa \equiv sb$ $\pmod{m}$ and $\gcd(s, m) = 1$, then $a \equiv b \pmod{m}$.*

Definition 1.5. *A* **prime** *number is an integer greater than 1 which has no (positive) factors except for itself and 1.*

Lemma 1.14. Characterisations of primes. *Let $p > 1$. Then the following are equivalent:*

- *p is prime;*
- *for all integers $m, n > 0$, if $p = mn$ then $m = 1$ or $n = 1$;*
- *for all integers m, n, if $p \mid mn$ then $p \mid m$ or $p \mid n$.*

We remark that in more general situations, the matter of primes is approached differently. The definition given above is, properly speaking, the definition of an *irreducible* rather than a prime, and the definition of a prime is the third property in the lemma. From this point of view, the lemma shows, essentially, that primes and irreducibles are the same in the integers. In extended number systems known as *integral domains* an irreducible need not be prime, though it is still true that a prime is necessarily irreducible: see, for example, [62], section 4.5.

Exercise. Let
$$S = \{\, a + b\sqrt{-5} \mid a, b \in \mathbb{Z} \,\}.$$

Show that in S the integers 2 and 3 are irreducible but not prime.

Theorem 1.15. The Fundamental Theorem of Arithmetic. *Every positive integer can be expressed in a unique way as a product of primes.*

Comment. We regard 1 as a "product" of *no* primes.

Definition 1.6. *Euler's function ϕ : for any positive integer m we define $\phi(m)$ to be the number of the integers $1, 2, \ldots, m$ which are relatively prime to m.*

Theorem 1.16. Fermat's "little" Theorem and Euler's Theorem.

- *Let p be prime. Then for any integer a, we have $a^p \equiv a \pmod{p}$; and if, in addition, a is not a multiple of p, then $a^{p-1} \equiv 1 \pmod{p}$.*

- *Let m be a positive integer and a an integer relatively prime to m. Then $a^{\phi(m)} \equiv 1 \pmod{m}$.*

Proof. To prove Euler's generalisation of Fermat's Theorem, write $s = \phi(m)$ and let $b_1, b_2, \ldots, b_s$ be those integers from $1, 2, \ldots, m$ which are coprime to m. Form the products $ab_1, ab_2, \ldots, ab_s$ and reduce them modulo m. The results are coprime to m, are in the range $1, 2, \ldots, m$ and are all distinct; therefore they are (in some order) the same as $b_1, b_2, \ldots, b_s$. Hence

$$b_1 b_2 \cdots b_s \equiv (ab_1)(ab_2) \cdots (ab_s) \pmod{m} \ ;$$

Euler's Theorem follows from this, and Fermat's Theorem is an easy consequence.

CHAPTER 2

Hermite's Method

Talk with Hermite... the more abstract
entities are to him like living creatures.

Henri Poincaré

A DEFECT OF CERTAIN TYPES of irrationality proof is that they apply largely, even perhaps solely, to artificially constructed examples. A case in point is the Champernowne constant

0.12345678910111213141516···

discussed in Chapter 1. In many ways a much more attractive problem is to investigate the irrationality of what might be called "naturally occurring" numbers. We have already seen (exercise 1.2) that $\sqrt{n}$ is irrational for positive integers n other than perfect squares; while no doubt "naturally occurring", these numbers form a somewhat limited class.

Johann Heinrich Lambert
(1728–1777)

We know too that e and e^2 are irrational (Theorem 1.9 and exercise 1.20): these facts were first proved, essentially, by Euler in 1737. Another obvious candidate for investigation is π. In fact, π is also irrational, a result originally due to J.H. Lambert ([37], English translation [54]) in 1761.

Lambert used relationships between infinite series and *continued fractions*, in particular, the formula

$$\tan\frac{s}{t} = \cfrac{s}{t - \cfrac{s^2}{3t - \cfrac{s^2}{5t - \cdots}}}$$

to prove that if r is a non–zero rational number then $\tan r$ is irrational. Equivalently, if $\tan r$ is rational and non–zero, then r is irrational, and it follows by taking $r = \frac{1}{4}\pi$ that π is irrational. By similar methods Lambert also showed that if r is rational and $r \neq 0$, then e^r is irrational. In Chapter 7 we shall study Lambert's approach to these questions, making use of arguments inspired by Lambert's use of continued fractions.

In the nineteenth century, Lambert's irrationality theorems were reconsidered by Charles Hermite, who succeeded in giving new proofs based upon entirely different ideas; Hermite's methods have turned out to be more valuable than his predecessor's. In fact, Hermite gave two related but distinct ways to attack such problems. One [31] involves what later became known as *Padé approximants*: quotients of polynomials that provide good approximations to the exponential function. The other [30] gives less information about the exponential function as a whole but simplifies the investigation of certain values of e^r. It is this second approach that we now discuss.

2.1   IRRATIONALITY OF e^r

In the actual details of the final proof, Hermite's method is (at least for the earlier results) not too difficult. However, the motivation behind the proof can be obscure. Therefore, instead of giving the proofs straight away, we shall start by trying to explain the aims and ideas behind a relatively simple case. We wish to generalise results of Chapter 1 by showing that if r is rational then e^r is irrational, with the obvious exception that $e^0 = 1$.

As usual we seek a proof by contradiction: take $r = a/b$ with $a \neq 0$, and suppose that $e^r = p/q$. Following the method of Theorem 1.9, we try to obtain a contradiction by constructing an integer that lies between 0 and 1. Hermite's idea, which originated in his study of approximations to e^x, was to consider the definite integral

$$\int_0^r f(x)\, e^x\, dx\,, \tag{2.1}$$

and to identify a function f which will give us what we want. Integrating by parts yields

$$\int_0^r f(x)\, e^x\, dx = \left(f(r)e^r - f(0)\right) - \int_0^r f'(x)\, e^x\, dx\,,$$

and since the integral on the right–hand side has very much the same form as that on the left, we may apply the same procedure repeatedly to obtain

$$\int_0^r f(x)\, e^x\, dx = \left(f(r) - f'(r) + f''(r) - \cdots\right)e^r - \left(f(0) - f'(0) + f''(0) - \cdots\right).$$

Here the right–hand side purports to contain two infinite series and therefore must be treated with caution, but if we choose f to be a polynomial, then the sums will actually involve a finite number of terms only, and we shall have no convergence problems. We write

$$F(x) = f(x) - f'(x) + f''(x) - f'''(x) + \cdots ,$$

so that

$$\int_0^r f(x) \, e^x \, dx = F(r) \, e^r - F(0) ,$$

and the next step is to make some sort of evaluation of the right–hand side. An idea that will help is to notice that $F(0)$ will be simple if f has a large number of derivatives that vanish at $x = 0$; that is, $f(x)$ should have many factors of x. Similarly, $f(x)$ should have many factors of $r - x$ in order to keep $F(r)$ simple. So we set

$$f(x) = cx^n(r - x)^n ,$$

where c, a constant, and n are yet to be chosen. Now

$$f^{(k)}(0) = k! \times \{ \text{coefficient of } x^k \}$$

$$= k! \, c \binom{n}{k - n} r^{2n-k}(-1)^{k-n} \tag{2.2}$$

if $n \le k \le 2n$, and $f^{(k)}(0) = 0$ otherwise. Recall that our aim is to make the integral (2.1), or something similar, an integer. The expression for $f^{(k)}(0)$ contains a factor r^{2n-k}, and this could have a denominator as big as b^n. Therefore, we choose $c = b^n$; then $f^{(k)}(0)$ is always an integer, and so is $F(0)$. Either by invoking the symmetry of f or by direct calculation, we note that

$$f(r - x) = f(x) \quad \Rightarrow \quad (-1)^k f^{(k)}(r - x) = f^{(k)}(x)$$

$$\Rightarrow \quad f^{(k)}(r) = (-1)^k f^{(k)}(0) .$$

Therefore, $F(r)$ is an integer too, and so is

$$q \int_0^r f(x) \, e^x \, dx = pF(r) - qF(0) . \tag{2.3}$$

We wish to show that this integer lies between 0 and 1, and thereby obtain a contradiction. Now the integrand on the left–hand side is the product of a (positive) exponential factor, and a polynomial which is zero at the points $x = 0$ and $x = r$, strictly positive in between. Therefore, the integral is strictly positive. Within the interval of integration, $f(x)$ is a constant times a product of $2n$ terms, each at most r; therefore

$$0 < q \int_0^r f(x) \, e^x \, dx < qrcr^{2n}e^r = pb^n r^{2n+1} .$$

This is clearly *not* small and our attempt has failed! But returning to (2.2), we notice that $f^{(k)}(0)$ always has a factor $n!$. Therefore, if we redefine c to be $b^n/n!$, then the expression (2.3) is still an integer and our estimate becomes

$$0 < q \int_0^r f(x) \, e^x \, dx < qrcr^{2n}e^r = pr \frac{(br^2)^n}{n!} .$$

Finally, remember that we have not yet chosen n. Note that p, r and b are fixed numbers, unchanged when n changes; and recall that if γ is any real constant, then $\gamma^n/n! \to 0$ as $n \to \infty$. Therefore, by choosing n large enough we obtain

$$0 < q \int_0^r f(x) \, e^x \, dx < 1 \ ,$$

and we have the desired contradiction.

Having seen the ideas behind Hermite's proof, let's now carefully write out the details: we shall give a variant of Hermite's proof due to Niven [44]. Since e^r is rational if and only if e^{-r} is rational, we may assume that r is positive (which was, in fact, tacitly assumed in the above working – where?). The point about evaluating the derivatives of f will be important in future proofs, so we isolate it in a lemma.

Lemma 2.1. Derivatives of polynomials. *Let a and b be integers, with $b \neq 0$. Let n be a non–negative integer; define the polynomial f by*

$$f(x) = (a - bx)^n g(x) \ ,$$

where g is a polynomial with integral coefficients.

- *If g has degree at most n, then for all $k \geq 0$ the derivative $f^{(k)}(a/b)$ is an integer divisible by $n!$.*

- *If $b = \pm 1$, the same conclusion holds, irrespective of the degree of g.*

Proof. The kth derivative of $(a - bx)^n$, evaluated at $x = a/b$, is zero unless $k = n$, in which case the derivative is $(-b)^n n!$. Differentiating f by Leibniz' formula

$$\frac{d^k(uv)}{dx^k} = \sum_{j=0}^k \binom{k}{j} \frac{d^j u}{dx^j} \frac{d^{k-j} v}{dx^{k-j}} \ ,$$

we see first that if $k < n$, then every term vanishes at $x = a/b$. If $k \geq n$, then the only surviving term is that with $j = n$, and so

$$f^{(k)}\left(\frac{a}{b}\right) = \binom{k}{n} (-1)^n \, n! \, b^n \, g^{(k-n)}\left(\frac{a}{b}\right) \ .$$

But since $g^{(k-n)}$ has integral coefficients, $g^{(k-n)}(a/b)$ is a rational number with denominator at most $b^{\deg g}$. So, if either $\deg g \leq n$ or $b = \pm 1$, then $b^n g^{(k-n)}(a/b)$ is an integer, and hence $f^{(k)}(a/b)$ is an integer times $n!$.

Theorem 2.2. (Lambert). *If r is a non–zero rational number, then e^r is irrational.*

Proof (Hermite/Niven). Let $r = a/b$ with $a \neq 0$, and suppose that $e^r = p/q$ is rational. As explained above, we may assume without loss of generality that r, a and b are positive. Now let

$$f(x) = \frac{x^n(a - bx)^n}{n!} \, ,$$

where n is a positive integer which will be specified later. Integrating by parts repeatedly, we have

Charles Hermite
(1822–1901)

$$\int_0^r f(x)\, e^x\, dx = \left(f(r)e^r - f(0)\right) - \int_0^r f'(x)\, e^x\, dx$$

$$= \left(f(r)e^r - f(0)\right) - \left(f'(r)e^r - f'(0)\right) + \int_0^r f''(x)\, e^x\, dx$$

$$= \cdots$$

$$= F(r)e^r - F(0) \, ,$$

where

$$F(x) = f(x) - f'(x) + f''(x) - \cdots \, .$$

Since f is a polynomial, the series is finite and there are no convergence problems. Lemma 2.1 (derivatives of polynomials) shows that $f^{(k)}(r)$ and $f^{(k)}(0)$ are always integers; thus

$$q \int_0^r f(x)\, e^x\, dx = pF(r) - qF(0) \tag{2.4}$$

is also an integer. However, on the interval $0 < x < r$ the polynomial $f(x)$ is always positive and has absolute value at most $a^n r^n / n!$; hence

$$0 < q \int_0^r f(x)\, e^x\, dx \leq qre^r \left(\frac{(ar)^n}{n!}\right) \, .$$

If n is chosen sufficiently large, the right–hand side is less than 1, which contradicts the fact that (2.4) is an integer; this proves the theorem.

A different combination of similar ingredients gives us an **alternative proof**. Again let $r = a/b$ be a non–zero rational number, and suppose that $e^r = p/q$. Define

$$I_n = q \int_0^r \frac{x^n(a - bx)^n}{n!} e^x\, dx = \frac{q}{n!} \int_0^r (ax - bx^2)^n e^x\, dx \, .$$

Then we have

$$I_0 = p - q \, , \quad I_1 = a(p + q) - 2b(p - q) \, ;$$

integrating by parts yields

$$I_{n+2} = q \int_0^r \left(\frac{-(4n+6)b(ax - bx^2)^{n+1}}{(n+1)!} + \frac{a^2(ax - bx^2)^n}{n!} \right) e^x \, dx$$

which can be written

$$I_{n+2} = -(4n+6)bI_{n+1} + a^2 I_n \ .$$

It is clear from this recurrence relation and the above initial conditions that I_n is always an integer. However, estimating the integral as in the previous proof gives the inequality $0 < |I_n| < 1$ for sufficiently large n, a contradiction. We invite the reader to fill in the details of this proof.

Corollary 2.3. Irrationality of logarithms. *If r is rational, positive and not equal to 1, then $\log r$ is irrational, where $\log$ denotes the natural logarithm of a positive real number.*

2.2   IRRATIONALITY OF π

Similar ideas can be used to prove the irrationality of π, and more generally, of $\cos r$ if r is rational and non–zero. Our aim will be to integrate $f(x) \sin x$ for suitable functions f; we shall find that the polynomial used in previous proofs is not always suitable.

Theorem 2.4. π *is irrational.*

Proof. Suppose that $\pi = a/b$; define

$$f(x) = \frac{(ax - bx^2)^n}{n!} = \frac{x^n(a - bx)^n}{n!}$$

and once again integrate by parts:

$$\int_0^\pi f(x) \sin x \, dx = \big(f(\pi) + f(0)\big) + \int_0^\pi f'(x) \cos x \, dx$$

$$= \big(f(\pi) + f(0)\big) - \int_0^\pi f''(x) \sin x \, dx \ .$$

In the second integration, we have used the fact that $\sin \pi = \sin 0 = 0$. Continuing to integrate in the same way we obtain

$$\int_0^\pi f(x) \sin x \, dx = \big(f(\pi) + f(0)\big) - \big(f''(\pi) + f''(0)\big) + \cdots$$

$$= F(\pi) + F(0) \ ,$$

where

$$F(x) = f(x) - f''(x) + f^{(iv)}(x) - f^{(vi)}(x) + \cdots \ .$$

Using Lemma 2.1 (derivatives of polynomials), and recalling that by assumption $\pi = a/b$, we see that both $F(\pi)$ and $F(0)$ are integers. But $f(x)\sin x$ is always positive for $0 < x < \pi$ and we have

$$0 < \int_0^\pi f(x)\sin x\, dx \leq \pi\,\frac{(a\pi)^n}{n!}\,.$$

If n is sufficiently large, the right–hand side is less than 1, and we have a contradiction in the usual manner. Therefore, π is irrational.

Comment. We might reasonably expect to obtain a similar proof by using the integral

$$\int_0^\pi f(x)\cos x\, dx$$

with the same polynomial $f(x)$. In fact, the attempt fails utterly! (*Exercise.* Explain why.) We can, however, prove the irrationality of π by considering

$$\int_{-\pi}^\pi \frac{(a - bx)^n(a + bx)^n}{n!}\cos x\, dx\;; \qquad (2.5)$$

though in this case, the integrand is not always positive for $-\pi < x < \pi$, and the fact that the integral is non–zero (while possibly "obvious" from figure 2.1) is slightly tricky to prove carefully.

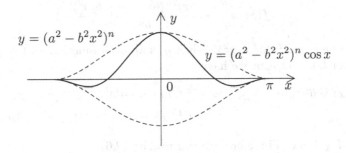

Figure 2.1 The graph of $y = (a^2 - b^2 x^2)^n \cos x$.

2.3 IRRATIONAL VALUES OF TRIGONOMETRIC FUNCTIONS

The proof of Theorem 2.4 can be viewed in a slightly different light. Taking $r = \pi$, we assumed that r is rational and used the fact that $\cos r$ is rational to reach a contradiction. However, the proof relied vitally on the fact that $\sin \pi = 0$, so one could not expect exactly the same proof to work for arbitrary rational r. Nevertheless, it turns out that by modifying the proof somewhat, we can prove the following result.

Theorem 2.5. Irrationality of cosines. *If r is rational and not zero, then $\cos r$ is irrational.*

Proof. Let $r = a/b$ be a non–zero rational; assume that $\cos r = p/q$. Without loss of generality assume that a and b, and hence r, are positive. Choose

$$f(x) = x^n(a - bx)^{2n}(2a - bx)^n \ ;$$

in this case we find it more convenient not to include $n!$ in the denominator. Integrating twice by parts yields

$$\int_0^r f(x)\sin x \, dx = f(0) - f(r)\cos r + f'(r)\sin r - \int_0^r f''(x)\sin x \, dx \ ;$$

repeating the procedure and writing

$$F(x) = f(x) - f''(x) + f^{(iv)}(x) - f^{(vi)}(x) + \cdots$$

gives eventually

$$\int_0^r f(x)\sin x \, dx = F(0) - F(r)\cos r + F'(r)\sin r \ . \tag{2.6}$$

Now observe that $f(x)$ is a polynomial in $(a - bx)^2$, since

$$f(x) = \frac{(a - bx)^{2n}(a^2 - (a - bx)^2)^n}{b^n} \ ;$$

if we set $g(x) = x^{2n}(a^2 - x^2)^n$, then g is an even function and so $g^{(k)}(0) = 0$ whenever k is odd. But we have

$$f(x) = b^{-n}g(a - bx) \quad \Rightarrow \quad f^{(k)}(x) = (-b)^k b^{-n}g^{(k)}(a - bx)$$
$$\Rightarrow \quad f^{(k)}(r) = -b^{k-n}g^{(k)}(0) = 0 \quad \text{for odd } k,$$

and so $F'(r) = 0$. Therefore, we can rewrite (2.6) as

$$q\int_0^r f(x)\sin x \, dx = qF(0) - pF(r) \ . \tag{2.7}$$

Now applying the lemma on derivatives of polynomials shows that $f^{(k)}(r)$ is a multiple of $(2n)!$, and hence also a multiple of $(n + 1)!$, for all k. In the case of $f^{(k)}(0)$, we need a little more information than is given by the lemma. Since

$$f^{(k)}(0) = k! \times \{ \text{coefficient of } x^k \}$$

we see that $f^{(k)}(0)$ is zero for $k < n$ and is divisible by $(n + 1)!$ for $k > n$. Moreover, for $k = n$ we have

$$f^{(n)}(0) = n!\, 2^n a^{3n} \ .$$

If $n+1$ is an odd prime greater than both a and q, we consider

$$qF(0) = \pm q\big(f^{(n)}(0) - f^{(n+2)}(0) + f^{(n+4)}(0) - \cdots\big) .$$

Every term on the right–hand side is a multiple of $n!$, and every term except the first is a multiple of $(n+1)!$. Hence $qF(0)$ is an integer which is a multiple of $n!$ but not of $(n+1)!$. We have seen above that $F(r)$ is a multiple of $(n+1)!$. Finally, therefore, the right–hand side of (2.7) is an integer divisible by $n!$, and not zero because it is not divisible by $(n+1)!$; so

$$|qF(0) - pF(r)| \geq n! .$$

The remainder of the proof follows familiar lines. By estimating the integral, we have

$$\left| q \int_0^r f(x) \sin x \, dx \right| \leq qr\, r^n a^{2n} (2a)^n = qr\, (2ra^3)^n ,$$

and for large n this is less than $n!$. So if $n+1$ is a sufficiently large prime (in particular, greater than a and q), then our estimates for the left and right–hand sides of (2.7) are incompatible. We have obtained a contradiction and thereby proved the theorem.

Comments. What are the features that make the above proof work? In particular, where did the choice of f come from? First, observe that we could have calculated the integral (2.6) *before* having specified the polynomial f. We would then have noted that the right–hand side contains a term $F'(r) \sin r$, and we know absolutely nothing about the factor $\sin r$. It would appear, then, that the only hope of success is to choose f in such a way that $F'(r)$ is zero. Since $F'(x)$ involves only odd-order derivatives of f, we give f a factor of $r-x$ to an even power, and "balance" the factor x^n with a factor $(2r-x)^n$, so that f is symmetrical about $x = r$. This yields the properties we need.

Note also the new method of proving that the expression (2.7) is not zero. We cannot do this, as before, by arguing that the integrand is positive, since the factor $\sin x$ might change sign frequently between 0 and r. Compare the graph in figure 2.1.

Corollary 2.6. Irrationality of sines and tangents. *If r is a non-zero rational number, then $\sin r$ and $\tan r$ are irrational.*

Proof. Let r be rational, $r \neq 0$. If $\sin r$ is rational, then by considering the double–angle formula

$$\cos 2r = 1 - 2\sin^2 r$$

we see that $2r$ is a non–zero rational with $\cos 2r$ rational. This contradicts the above result. The proof for $\tan r$ is similar, using the formula

$$\cos 2r = \frac{\cos^2 r - \sin^2 r}{\cos^2 r + \sin^2 r} = \frac{1 - \tan^2 r}{1 + \tan^2 r} .$$

EXERCISES

2.1 Let

$$I_n = \int_{-1}^{1} (1 - x^2)^n \cos \pi x \, dx \ .$$

Find a reduction formula giving I_{n+2} in terms of I_{n+1} and I_n, and use it to show that π^2 is irrational.

2.2 Prove that the area under the graph in figure 2.1 is not zero.

2.3 By considering the integral

$$I = \int_0^{\pi} f(x) \sin x \, dx$$

for a suitable polynomial $f(x)$, show that if c is a positive integer then $\pi \sqrt{c}$ is irrational. Hence give a one–line proof of the irrationality of π^2.

2.4 Let r be rational and not zero. Show that $\cosh r$ is irrational, and deduce that $\sinh r$, $\tanh r$ and e^r are irrational.

2.5 By considering

$$I = \int_0^1 f(x) \sin rx \, dx \quad \text{where} \quad f(x) = x^n (1 - x)^{2n} (2 - x)^n \ ,$$

prove that if r^2 is a non–zero rational number, then $\cos r$ is irrational.

2.6 The Bessel function of the first kind of order ν, defined by

$$J_\nu(x) = \sum_{k=0}^{\infty} \frac{(-1)^k}{k! \, \Gamma(\nu + k + 1)} \left(\frac{x}{2} \right)^{2k+\nu} \ ,$$

is encountered in the solutions of many important differential equations. It has the property

$$x J_{\nu+1}(x) = 2\nu J_\nu(x) - x J_{\nu-1}(x) \ ,$$

which is easy to prove, and

$$J_\nu(x) = \frac{1}{\sqrt{\pi} \, \Gamma(\nu + \frac{1}{2})} \left(\frac{x}{2} \right)^\nu \int_0^{\pi} \cos(x \cos \theta) \sin^{2\nu} \theta \, d\theta \quad \text{for } \nu > -\tfrac{1}{2},$$

which is not. The gamma function, which appears in the above formulae, is defined by

$$\Gamma(x) = \int_0^{\infty} t^{x-1} e^{-t} \, dt$$

for real $x > 0$. It is increasing for $x \geq 1$ and satisfies $\Gamma(n+1) = n!$ when n is a non–negative integer.

(a) Let $r^2 = a/b$ and $\nu = c/d$, and suppose that $rJ_{\nu+1}(r)/J_\nu(r) = p/q$. Prove that
$$r^n J_{\nu+n}(r) = ru_n J_{\nu+1}(r) + v_n J_\nu(r)$$
for $n \geq 0$, where the sequences u_n and v_n are defined by
$$u_{n+1} = 2(\nu + n)u_n - r^2 u_{n-1}, \quad v_{n+1} = 2(\nu + n)v_n - r^2 v_{n-1}$$
with suitable initial conditions.

(b) Show that if n is an integer, $n \geq 0$, then $qb^n d^n r^n J_{\nu+n}(r)/J_\nu(r)$ is an integer.

(c) Let t be an integer greater than $-\nu + \frac{1}{2}$. For sufficiently large integers n, obtain contradictory estimates for
$$\left| \frac{qb^{n+t}d^{n+t}r^{n+t}J_{\nu+n+t}(r)}{J_\nu(r)} \right|,$$
and hence. . .

(d) . . . conclude that if ν is rational, r^2 is rational and not zero, and $J_\nu(r) \neq 0$, then
$$\frac{rJ_{\nu+1}(r)}{J_\nu(r)}$$
is irrational.

(e) The gamma function also has the property that $\Gamma(x + 1) = x\Gamma(x)$ for all $x > 0$, including non–integer x. Use this to deduce that under suitable conditions $r \tan r$ is irrational.

2.7 *For readers who know some elementary complex analysis.* We follow a method of Desbrow [22] to prove that if r^2 is a non–zero rational number, then $r \tanh r$ is irrational. Write
$$f(z) = \sum_{n=0}^{\infty} \frac{r^{2n} z^n}{(2n)!}$$
and assume that $r^2 = a/b$ and $r \tanh r = p/q$ are both rational.

(a) Explain why f has a Taylor series
$$f(z) = \sum_{n=0}^{\infty} c_n (z - 1)^n$$
valid for all $z \in \mathbb{C}$. Show that the coefficients of the series, except for c_0, can be found by integrating around a circle C in the complex plane with centre 1 and radius n^2 to give
$$c_n = \frac{1}{2\pi i} \int_C \frac{f(z)}{(z - 1)^{n+1}} \, dz .$$

(b) Show that $w = f(z)$ is a solution of the differential equation

$$4zw'' + 2w' - r^2 w = 0 ,$$

and hence find a second–order recurrence for the coefficients c_n. Evaluate c_0 and c_1.

(c) Prove that $d_n = n!\, 2^n b^n q\, c_n/c_0$ is an integer for all n and is non–zero for infinitely many n.

(d) Show that $|f(z)| \le e^{r|z|^{1/2}}$ for all complex z.

(e) Use these results to obtain a contradiction and so prove the result claimed.

APPENDIX: SOME RESULTS OF ELEMENTARY CALCULUS

Proofs of the following results are again left up to the reader. If assistance is required, any basic calculus text should suffice.

Lemma 2.7. *For any real constants β and γ, we have*

$$\frac{\beta \gamma^n}{n!} \to 0 \quad as\ n \to \infty .$$

Corollary 2.8. Comparison of exponentials and factorials. *If β and γ are real constants, then $|\beta \gamma^n| < n!$ for all sufficiently large integers n.*

Lemma 2.9. Derivatives of even and odd functions. *Let f be a differentiable function from $\mathbb{R}$ to $\mathbb{R}$. Then*

- *f is even if and only if f' is odd;*
- *if f is odd then f' is even.*

Exercise. Why is the converse of the second result not true? Fill in the gap and then prove the statement:

"if f' is even and … then f is odd".

Corollary 2.10. *If g is an odd function, then $g^{(k)}(0) = 0$ for all even k. If g is an even function, then $g^{(k)}(0) = 0$ for all odd k.*

Lemma 2.11. Derivatives of a product (Leibniz' rule). *If $k \ge 0$, then*

$$\frac{d^k(uv)}{dx^k} = \sum_{j=0}^{k} \binom{k}{j} \frac{d^j u}{dx^j} \frac{d^{k-j} v}{dx^{k-j}} .$$

Lemma 2.12. Estimation of integrals. *If $|\phi(x)| \leq M$ for all $x \in [a,b]$, and if the integral exists, then*

$$\left| \int_a^b \phi(x)\, dx \right| \leq M\, |b - a| \; .$$

In particular,

$$\left| \int_a^b \phi(x)\, dx \right| \leq |b - a| \max_{a \leq x \leq b} |\phi(x)|$$

provided the maximum exists.

CHAPTER 3

Algebraic and Transcendental Numbers

*The meaning doesn't matter
if it's only idle chatter
of a transcendental kind.*

W.S. Gilbert

A RATIONAL NUMBER is one which is the root of a linear polynomial $qz - p$ with integer coefficients. The next level of complexity in the arithmetic properties of real and complex numbers is reached by considering roots of polynomials of higher degree, still having integer (or, equivalently, rational) coefficients. The reader may care to ponder whether or not it is obvious that a complex number need not be a root of any such polynomial.

3.1 DEFINITIONS AND BASIC PROPERTIES

We begin with a definition which was given in a slightly different, but equivalent, form in Chapter 1.

Definition 3.1. *If the complex number α is a root of a polynomial*

$$a_n z^n + a_{n-1} z^{n-1} + \cdots + a_1 z + a_0 \qquad (3.1)$$

*with rational coefficients and $a_n \neq 0$, then α is said to be **algebraic**. Any (complex) number which is not algebraic is called a **transcendental** number. More generally, if α is a root of a polynomial such as (3.1) with coefficients in a field $\mathbb{F}$, then α is said to be **algebraic over** $\mathbb{F}$; if there is no such polynomial then α is **transcendental over** $\mathbb{F}$.*

DOI: 10.1201/9781003111207-3

Lemma 3.1. Properties of algebraic numbers.

- *A complex number α is algebraic (over $\mathbb{Q}$) if and only if it is a root of a non–zero polynomial with* integral *coefficients.*

- *If α is algebraic, then there exists a unique monic polynomial f_α having rational coefficients and smallest possible degree, such that $f_\alpha(\alpha) = 0$. If g is any polynomial with rational coefficients such that $g(\alpha) = 0$, then g is a multiple of f_α.*

- *The polynomial f_α is irreducible over $\mathbb{Q}$. That is, f_α cannot be factorised as the product of two polynomials with rational coefficients and degree smaller than that of f_α.*

Proof. To prove the first assertion, just multiply a polynomial with rational coefficients by a common denominator for its coefficients. In the second statement the existence of f_α is clear; if $g(\alpha) = 0$ then dividing g by f_α gives

$$g(z) = f_\alpha(z)q(z) + r(z)$$

with $r(\alpha) = 0$. But r has smaller degree than f_α, so r is the zero polynomial and hence g is a multiple of f_α. The uniqueness of f_α follows since if there are two polynomials with the minimal–degree property then each is a factor of the other. To prove irreducibility note that if there is a proper factorisation $f_\alpha = gh$ then either g or h has α as a root, thus contradicting the minimality of f_α.

Definition 3.2. *The polynomial f_α of the above lemma is called the* **minimal polynomial** *of α. The* **degree** *of an algebraic number is the degree of its minimal polynomial. If the minimal polynomial of α has integral coefficients, then α is called an* **algebraic integer**. *Algebraic numbers having the same minimal polynomial are said to be* **conjugate** *to each other.*

Lemma 3.2. Gauss' Lemma. *Let f be a polynomial with integral coefficients and suppose that $f = gh$, where g and h are polynomials with rational coefficients. Then there is a rational constant c such that cg and h/c have integral coefficients, and so f is a product of polynomials with integral coefficients and the same degrees as g and h.*

Proof. First, observe that if the polynomial $\overline{g}$ has integer coefficients with no common factor, and if $\overline{h}$ also has integer coefficients with no common factor, then the same is true of the product $\overline{g}\overline{h}$. For let p be a prime and suppose that p divides $g_0, g_1, \ldots, g_{k-1}$ but not g_k; and that p divides $h_0, h_1, \ldots, h_{\ell-1}$ but not h_ℓ. Then p is not a factor of

$$g_0 h_{k+\ell} + g_1 h_{k+\ell-1} + \cdots + g_{k-1}h_{\ell+1} + g_k h_\ell + g_{k+1}h_{\ell-1} + \cdots + g_{k+\ell}h_0 \,,$$

which is one of the coefficients of $\overline{g}\overline{h}$. Thus there is no prime which divides *every* coefficient of $\overline{g}\overline{h}$. To prove Gauss' Lemma, let $f = gh$ where g and h

have rational coefficients. Consider the coefficients of g: by extracting the least common multiple of the denominators and then the greatest common divisor of the numerators, we can write $g = (s/t)\overline{g}$, where $\overline{g}$ has integral coefficients with no common factor; similarly, $h = (u/v)\overline{h}$. Thus

$$tv\, f = su\, \overline{g}\overline{h} \ .$$

Since the coefficients of $\overline{g}\overline{h}$ have no common factor, su is a multiple of tv and so $(s/t)(u/v)\overline{g}$ has integral coefficients. The lemma follows upon choosing $c = u/v$.

Corollary 3.3. *If α is a root of any monic polynomial with integral coefficients, then α is an algebraic integer.*

Comment. It follows from this corollary that the definition we have just given of an algebraic integer is equivalent to the (slightly different) definition we gave in Chapter 1.

Examples.

- The polynomial $z^2 + 3$ is clearly irreducible over $\mathbb{Q}$. It has two roots $\pm i\sqrt{3}$, each of which is, therefore, an algebraic integer of degree 2.

- By using multiple angle formulae, or otherwise, it is possible to show that $\cos \frac{1}{7}\pi$, $\cos \frac{3}{7}\pi$ and $\cos \frac{5}{7}\pi$ are the roots of the cubic $8z^3 - 4z^2 - 4z + 1$. This polynomial is irreducible, so the roots are algebraic of degree 3, and are conjugate to each other. Note that in this case, conjugate algebraic numbers are *not* complex conjugates.

- The fifth root of unity $\zeta = e^{2\pi i/5}$ is a root of $z^5 - 1$. This polynomial is not irreducible since

$$z^5 - 1 = (z - 1)(z^4 + z^3 + z^2 + z + 1) \ ,$$

but the quartic factor then is irreducible. (*Proof.* The quartic has no rational roots and therefore no linear factor; if, therefore, it is reducible then it is the product of two quadratics. By the above lemma we may assume the factors have integral coefficients; multiplying out and equating coefficients we soon obtain a contradiction.) Therefore, ζ is an algebraic integer of degree 4. Its conjugates are ζ^2, ζ^3 and ζ^4.

- The numbers e and π are both transcendental over $\mathbb{Q}$. We'll give proofs in Chapter 5.

3.1.1   Proving polynomials irreducible

The next two results are often useful for proving irreducibility of polynomials.

Lemma 3.4. Eisenstein's Lemma. *Let f be a polynomial with integral coefficients,*

$$f(z) = a_n z^n + a_{n-1} z^{n-1} + \cdots + a_1 z + a_0 \ ;$$

suppose that there is a prime p such that p is a factor of $a_0, a_1, \ldots, a_{n-1}$ but not of a_n, and such that p^2 is not a factor of a_0. Then f is irreducible over the field $\mathbb{Q}$ of rational numbers.

Proof. By Gauss' Lemma, we need only show that f cannot be written as the product of two polynomials with integer coefficients and degree less than n. Suppose that there are two such polynomials; without loss of generality we may assume that they have the same number of terms, say

$$f(z) = g(z)h(z) = (b_m z^m + \cdots + b_1 z + b_0)(c_m z^m + \cdots + c_1 z + c_0)$$

with $m < n$. Looking at the constant terms, $b_0 c_0 = a_0$ is divisible by p but not by p^2; by symmetry, we may assume that $p \mid b_0$ and $p \nmid c_0$. Multiplying out all the other coefficients shows that p is a factor of $b_1, b_2, \ldots, b_m$ too. But then p is a factor of every a_k, including a_n, and this is contrary to our initial assumption. So f is irreducible.

Examples.

- The polynomial $z^3 - 12z^2 + 345z - 6789$ is irreducible since 3 is a prime factor of 12, of 345 and of 6789 but not of the leading coefficient, and since 3^2 is not a factor of 6789.

- A slightly more subtle application of Eisenstein's Lemma simplifies the proof of irreducibility for

$$f(z) = \frac{z^5 - 1}{z - 1} = z^4 + z^3 + z^2 + z + 1 \ .$$

It is not hard to see that we can factorise $f(z)$ if and only if we can factorise

$$f(z + 1) = \frac{(z + 1)^5 - 1}{z} = z^4 + 5z^3 + 10z^2 + 10z + 5 \ ;$$

but Eisenstein's Lemma with $p = 5$ shows immediately that $f(z + 1)$ is irreducible, and hence so is $f(z)$.

- In fact, if p is any prime then $f(z) = z^{p-1} + z^{p-2} + \cdots + z + 1$ can be proved irreducible by the same method. **Exercise.** Give the details!

Notation. By $\mathbb{Z}_m$ we denote the ring $\{0, 1, 2, \ldots, m-1\}$ of integers modulo a positive integer m. Whenever we work in this ring it is to be understood that addition, subtraction and multiplication are performed modulo m. Any reader requiring a brief review of modular (congruence) arithmetic should consult the appendix to Chapter 1.

Lemma 3.5. Factorisation of polynomials modulo m. *Let f be a polynomial with integer coefficients and suppose that m is not a factor of its leading coefficient. Write f_m for the polynomial obtained by reducing the coefficients of f modulo m. If f has a factor of degree n over $\mathbb{Z}$, then f_m has a factor of degree n over $\mathbb{Z}_m$.*

Sketch of proof. Suppose that $f = gh$, where g has degree n. Since equality of integers implies congruence to any modulus, we have $f_m = g_m h_m$. Moreover, g_m cannot have degree greater than n, and will have degree less than n only if m is a factor of the leading coefficient of g; but then m would be a factor of the leading coefficient of f, contrary to assumption. Thus g_m has degree n.

Corollary 3.6. Testing reducibility with modular arithmetic. *If f, a polynomial with integral coefficients, is reducible over $\mathbb{Z}$, and if m is not a factor of its leading coefficient, then f_m is reducible over $\mathbb{Z}_m$.*

We shall commonly apply this corollary in contrapositive form: if there is any m, not a factor of the leading coefficient of f, for which f_m is irreducible, then f is irreducible. This can be a good way to prove a function irreducible, as there is only a limited number of irreducible polynomials with coefficients in $\mathbb{Z}_m$; however, finding a suitable modulus m may require a little ingenuity – or a lot of trial and error!

Examples.

- Let $f(z) = z^3 - 4z^2 + 9z + 16$. Choose $m = 3$; we have

$$f_3(z) = z^3 + 2z^2 + 1 .$$

If f_3 is reducible it must have a linear factor. However, we can easily calculate in $\mathbb{Z}_3$ that

$$f_3(0) = 1 , \quad f_3(1) = 1 \quad \text{and} \quad f_3(2) = 2 ;$$

so f_3 has no roots in $\mathbb{Z}_3$, hence no linear factors, and therefore is irreducible. By the above corollary, f is also irreducible.

- Let $f(z) = 3z^5 + 11z^4 - 12z^3 + 6z - 21$ and try $m = 2$. We have

$$f_2(z) = z^5 + z^4 + 1$$

and $f_2(0) = f_2(1) = 1$. So f_2 has no linear factors, and nor does f. Being of degree 5, however, f_2 could be the product of a quadratic and a cubic; in fact, it is not too hard to discover that

$$f_2(z) = (z^2 + z + 1)(z^3 + z + 1) .$$

Thus f_2 is reducible; but note that the converse of the above result is not true, and so it does not apply to this example.

- Try the previous example with $m = 5$. We know from our previous attempt that f has no linear factors, and so the only potential factorisation that we need to investigate modulo 5 is

$$f_5(z) = 3z^5 + z^4 + 3z^3 + z + 4 = (az^2 + bz + c)(dz^3 + ez^2 + fz + g) .$$

We may assume (why?) that $a = 1$ and $d = 3$; multiplying out and equating coefficients gives the simultaneous equations

$$1 = 3b + e, \ 3 = f + be + 3c, \ 0 = g + bf + ce, \ 1 = bg + cf, \ 4 = cg$$

in $\mathbb{Z}_5$. As each coefficient takes one of only five possible values it is not excruciatingly difficult to try them all.

- If $b = 1$ the first three equations give $e = 3$, $f = 2c$ and $g = 0$. But then the last equation says $4 = 0$, which is obviously impossible.
- If $b = 2$ we obtain $e = 0$, $f = 3 + 2c$ and $g = 4 + c$. Now the final equation gives

$$4 = cg = 4c + c^2 \quad \Rightarrow \quad c^2 + 4c + 4 = 3 \quad \Rightarrow \quad (c + 2)^2 = 3 \ ;$$

but this is impossible as the squares modulo 5 are 0, 1 and 4 only.

Exercise. Rule out the remaining cases and hence prove that f is irreducible.

- The polynomial $f(z) = 2z^2 + 3z + 1$ is reducible over $\mathbb{Z}$; but $f_2(z) = z + 1$ is irreducible. This shows that we cannot neglect the requirement that m be not a factor of the leading coefficient of f in the above result.

Comments.

- The integers modulo a prime form a field, and therefore have a "nicer" arithmetic than the integers to a composite modulus. For this reason it is customary (though not obligatory) to take m prime in the above type of problem. However, sometimes a composite modulus will work when prime moduli don't – see exercise 3.13.

- In a composite modulus the Factor Theorem for polynomials must be used with caution. The theorem as usually stated is still true –

  "α is a root of f if and only if $z - \alpha$ is a factor of $f(z)$"

  – but if f has no roots it still may have a linear factor $az + b$ with, necessarily, $a \neq 1$. For a specific example, consider the polynomial $f(z) = 3z^2 + 3z + 2$. It is easy to see that for any integer α, we have $f(\alpha) \equiv 2 \pmod{6}$, and so f_6 has no roots in $\mathbb{Z}_6$; but we cannot conclude from this that f_6 is irreducible, and in fact $f_6(z) = (3z + 1)(3z + 2)$.

Lemma 3.7. Denominator of an algebraic number. *If α is an algebraic number, then there is a rational integer $d \neq 0$ such that $d\alpha$ is an algebraic integer.*

Proof. Suppose that

$$a_n \alpha^n + a_{n-1} \alpha^{n-1} + a_{n-2} \alpha^{n-2} + \cdots + a_1 \alpha + a_0 = 0 \ ,$$

where each a_k is an integer and $a_n \neq 0$. Choose $d = a_n$: multiplying both sides of the above equation by a_n^{n-1} and rearranging, we have

$$(a_n\alpha)^n + a_{n-1}(a_n\alpha)^{n-1} + a_{n-2}a_n(a_n\alpha)^{n-2} + \cdots + a_1a_n^{n-2}(a_n\alpha) + a_0a_n^{n-1} = 0 .$$

Thus $a_n\alpha$ is a root of a monic polynomial with integral coefficients, and by Corollary 3.3 is an algebraic integer.

Definition 3.3. *For any algebraic number α, the smallest positive integer d such that $d\alpha$ is an algebraic integer is called the **denominator** of α and is denoted* den α.

Examples.

- If $\alpha = p/q$ is a rational number, with p and q having no common factor and $q > 0$, then den $\alpha = q$. That is, den α is just the denominator, in the usual sense, of a fraction.

- Let $\alpha = \cos \frac{1}{7}\pi$. Then $8\alpha^3 - 4\alpha^2 - 4\alpha + 1 = 0$, as on page 33, and the proof of the lemma shows that 8α is an algebraic integer. However, 8 is not the smallest integer with this property, since

$$(2\alpha)^3 - (2\alpha)^2 - 2(2\alpha) + 1 = 0 .$$

It is not hard to see that α itself is not an algebraic integer; therefore $d = 1$ is not possible and we have den $\alpha = 2$.

3.1.2 Closure properties of algebraic numbers

Next we shall sketch proofs that the set of (complex) algebraic numbers forms a subfield of $\mathbb{C}$, and that the algebraic integers form an integral domain. These proofs require a certain acquaintance with basic properties of vector spaces and abelian groups; however, the level required is probably too much to summarise in an appendix. Therefore, on this occasion only, we invite the interested reader to refer to other sources for background material. Two of many possibilities are Axler [8] for linear algebra, Stewart and Tall [62] for groups. The reader who prefers to continue with the main topics of this book can safely proceed to section 3.2 after noting carefully the results of Theorem 3.10, Corollary 3.11 and Theorem 3.12.

Lemma 3.8. *Let $S = \{ \alpha_k \mid k \in K \}$ be a set of complex numbers. Then*

- *the set of linear combinations*

$$\sum r_k\alpha_k$$

with finitely many terms and rational coefficients r_k is a vector space over the field $\mathbb{Q}$;

- *the set of linear combinations*

$$\sum m_k \alpha_k$$

with finitely many terms *and integer coefficients* m_k *is an abelian group under addition.*

Lemma 3.9. Finiteness criteria for algebraic numbers. *Let* $\alpha \in \mathbb{C}$; *in the previous lemma take* $S = \{1, \alpha, \alpha^2, \dots\}$. *Then*

- α *is algebraic if and only if the vector space of rational linear combinations of* S *is finite–dimensional;*

- α *is an algebraic integer if and only if the group of integer linear combinations of* S *is finitely generated.*

Sketch of proof. If α is algebraic of degree n, then all powers of α can be written as linear combinations of $\{1, \alpha, \alpha^2, \dots, \alpha^{n-1}\}$, so the vector space has a spanning set (in fact, a basis) with n elements, and so is finite–dimensional. Conversely, if the vector space has dimension n, then $\{1, \alpha, \alpha^2, \dots, \alpha^n\}$ is a linearly dependent set, and this yields a polynomial identity satisfied by α.

If α is an algebraic integer of degree n, then every power of α can be written as an *integral* linear combination of $\{1, \alpha, \alpha^2, \dots, \alpha^{n-1}\}$, and so this set generates the group. Conversely, suppose that the group is generated by n elements $p_1, p_2, \dots, p_n$. Each of these is an integer linear combination of powers of α; therefore so are $\alpha p_1, \alpha p_2, \dots, \alpha p_n$, and we can write equations

$$\alpha p_k = m_{k1} p_1 + m_{k2} p_2 + \dots + m_{kn} p_n \quad \text{for} \quad k = 1, 2, \dots, n .$$

Transferring all the terms $m_{kj} p_j$ to the left–hand side gives a homogeneous system of linear equations with a non–zero solution; therefore the determinant

$$\begin{vmatrix} \alpha - m_{11} & -m_{12} & \cdots & -m_{1n} \\ -m_{21} & \alpha - m_{22} & \cdots & -m_{2n} \\ \vdots & \vdots & \ddots & \vdots \\ -m_{n1} & -m_{n2} & \cdots & \alpha - m_{nn} \end{vmatrix}$$

is zero. Expanding the determinant gives a monic polynomial in α with integral coefficients.

Theorem 3.10. Sums, differences and products of algebraic numbers. *If* α *and* β *are algebraic, then so are* $\alpha \pm \beta$ *and* $\alpha\beta$. *If* α *and* β *are algebraic integers, then so are* $\alpha \pm \beta$ *and* $\alpha\beta$.

Sketch of proof. This is an application of the previous lemma. If α is algebraic of degree m, then every power of α can be expressed in terms of the powers $1, \alpha, \alpha^2, \ldots, \alpha^{m-1}$; a similar observation holds for an algebraic number β of degree n. But every power of $\alpha + \beta$ can be written as

$$(\alpha + \beta)^k = \sum_{j=0}^{k} \binom{k}{j} \alpha^j \beta^{k-j} ;$$

expressing each α^j and β^{k-j} in terms of the lowest possible powers, expanding and collecting terms gives a result of the form

$$(\alpha + \beta)^k = \sum_{i=0}^{m-1} \sum_{j=0}^{n-1} r_{ij} \alpha^i \beta^j .$$

Therefore, the vector space generated by the powers of $\alpha + \beta$ is also spanned by the finite set $\{ \alpha^i \beta^j \mid 0 \leq i < m, 0 \leq j < n \}$, and so $\alpha + \beta$ is algebraic. The proofs of all the other assertions are similar.

Comment. The above proof also shows that the degrees of $\alpha \pm \beta$ and of $\alpha\beta$ are at most the degree of α times the degree of β.

Corollary 3.11. Quotients of algebraic numbers. *If α and β are algebraic and $\beta \neq 0$, then α/β is algebraic.*

Proof. All we need show is that a non–zero algebraic number has an algebraic reciprocal. But if β is a root of

$$b_n z^n + b_{n-1} z^{n-1} + \cdots + b_1 z + b_0 ,$$

then β^{-1} is a root of

$$b_0 z^n + b_1 z^{n-1} + \cdots + b_{n-1} z + b_n ;$$

this proves the result.

A result that we mentioned in Chapter 1 is proved by methods similar to those of the last theorem.

Theorem 3.12. Polynomials with algebraic coefficients. *If the complex number β is a root of the polynomial*

$$\alpha_n z^n + \alpha_{n-1} z^{n-1} + \cdots + \alpha_1 z + \alpha_0 \tag{3.2}$$

with algebraic coefficients and $\alpha_n \neq 0$, then β is algebraic. If β is a root of the monic polynomial

$$z^n + \alpha_{n-1} z^{n-1} + \cdots + \alpha_1 z + \alpha_0$$

whose coefficients are algebraic integers, then β is an algebraic integer.

Sketch of proof. Since $\alpha_n \neq 0$ and we know that quotients of algebraic numbers are algebraic, we may assume that $\alpha_n = 1$ in (3.2). Then every power of β can be written as a rational linear combination of expressions

$$\alpha_0^{m_0} \alpha_1^{m_1} \cdots \alpha_{n-1}^{m_{n-1}} \beta^m \qquad (3.3)$$

with $0 \leq m < n$. But by assumption, each α_k is algebraic of degree (say) d_k, and so can be written as a linear combination of $1, \alpha_k, \alpha_k^2, \ldots, \alpha_k^{d_k-1}$. Substituting these linear combinations into the linear combination (3.3) and expanding shows that every power of β is a linear combination of expressions like (3.3) in which the exponents satisfy

$$0 \leq m_k < d_k \text{ for all } k \quad \text{and} \quad 0 \leq m < n \ .$$

There are only finitely many such expressions (in fact, at most $d_0 d_1 \cdots d_{n-1} n$), so the vector space spanned by the powers of β is finite–dimensional, and β is algebraic.

To prove the assertion involving algebraic integers, just observe that all the linear combinations and all the expansions we have considered above will have integral coefficients.

3.2 EXISTENCE OF TRANSCENDENTAL NUMBERS

The first question we need to address about transcendental numbers is whether or not there are any! It is clear that algebraic numbers exist: for a start, all rational numbers are algebraic, and we have also given a few examples of irrational algebraic numbers. However, it is conceivable that *every* complex number could be a root of a rational polynomial, in which case transcendental numbers would not exist.

Notice, by the way, that we have so far only seen algebraic numbers of degree up to 4. It is not at all clear that algebraic numbers of arbitrarily high degree exist. If, for example, we were to consider polynomials with real (rather than rational) coefficients, then there would be no irreducible polynomials of degree greater than 2. The situation in this case would therefore be very simple: all real numbers would be algebraic (over $\mathbb{R}$) of degree 1, and all non–real complex numbers would be algebraic (over $\mathbb{R}$) of degree 2. Among the complex numbers there would be no algebraic numbers of higher degree, and no transcendental numbers.

The existence of transcendental numbers was first proved by Joseph Liouville, who attempted to show that e is not an algebraic number. He failed in this aim but achieved enough to allow him in 1844 (and again, using different techniques, in 1851) to give specific examples of transcendental numbers. A completely different proof was given three decades later by Georg Cantor: a proof which is perhaps simpler, though, as it does not provide any specific examples of transcendentals, possibly somehow beside the point as far as number theory is concerned. We shall begin with Cantor's proof.

Cantor proved the existence of transcendental numbers simply by showing that there are, in a sense, more complex numbers than algebraic numbers. Specifically, the set of complex numbers is uncountable – this follows immediately from the uncountability of the reals, proved by Cantor in 1874 – while, as we shall now show, the set of (complex) algebraic numbers is countable.

First, a slightly informal proof. Recall that an algebraic number is, (almost) by definition, a root of a non–zero polynomial with integral coefficients. Define the **height** of any such polynomial to be the maximum of the absolute values of its coefficients: that is, if $f(z) = a_n z^n + a_{n-1} z^{n-1} + \cdots + a_1 z + a_0$ with all a_k integers and $a_n \neq 0$, then

$$H(f) = \max\big(|a_n|, |a_{n-1}|, \ldots, |a_1|, |a_0|\big) \ .$$

The height of any polynomial $f \neq 0$ is a positive integer. Let m be a positive integer and consider all polynomials f with $\deg f + H(f) = m$. Any such polynomial has degree less than m and therefore at most m non–zero coefficients; each of these coefficients is an integer from $-m$ to m. Therefore, the number of f satisfying $\deg f + H(f) = m$ is at most $(2m + 1)^m$ and is hence finite. So we can construct a list of all non–zero polynomials over $\mathbb{Z}$ by writing down those with degree plus height equal to 1, followed by those with degree plus height equal to 2, and so on. We can now write down all the roots of the first polynomial in our list, followed by all the roots of the second, and so forth; deleting any repetitions in this list, we have listed all (real and complex) algebraic numbers once each. Since the set of algebraic numbers can be arranged in a list it is countable, and therefore cannot include all complex numbers. Thus transcendental numbers exist.

Theorem 3.13. *Transcendental numbers exist.*

Proof. We give a more formal version of the proof outlined in the previous paragraph. See appendix 1 at the end of this chapter for basic results on countability. Let $\mathbb{Z}[z]$ be the set of polynomials in z with integral coefficients, and let S be the set of finite sequences of integers. Since $\mathbb{Z}$ is countable, S is also countable. It is easy to see that, taking $a_n \neq 0$ as usual,

$$a_n z^n + a_{n-1} z^{n-1} + \cdots + a_1 z + a_0 \mapsto (a_n, a_{n-1}, \ldots, a_1, a_0)$$

defines a one–to–one function from $\mathbb{Z}[z]$ to S, and therefore $\mathbb{Z}[z]$ is countable. For each polynomial f in $\mathbb{Z}[z]$ let S_f be the set of roots of f. Then

$$\{\,\text{algebraic numbers}\,\} = \bigcup_{f \in \mathbb{Z}[z]} S_f$$

is countable by a result from the appendix. Since $\mathbb{C}$ is not countable, transcendental numbers exist.

Comment. According to the *intuitionist* school in the philosophy of mathematics (originating with L.E.J. Brouwer, 1881–1966), an existence proof is not valid unless it explicitly provides an algorithm for the construction of the object whose existence is asserted. In particular, the above proof would not be accepted by those who adopt this stance. The same objection is made to the ε–δ style of proof in elementary analysis; intuitionists have to reformulate definitions and then either re-prove or abandon results concerning limiting processes. A great deal of work has been done on this task; for an introduction see Körner [36], Chapters VI and VII, or for basic calculus from an intuitionist point of view try Bishop [15].

3.3 APPROXIMATION OF REAL NUMBERS BY RATIONALS

Instead of relying on Cantor's countability argument we can go back to Liouville's earlier proofs. These make it possible to explicitly construct transcendental numbers and are therefore, from the number–theoretic point of view, more interesting than Cantor's proof. They also avoid the intuitionist objections mentioned above – though perhaps not completely so, as they do make use of the Mean Value Theorem from elementary calculus.

Joseph Liouville
(1809–1882)

Liouville's methods derive from an investigation of the problem of **approximating real numbers by rationals**. Let $\alpha \in \mathbb{R}$; we wish to ask how closely α can be approximated by rational numbers p/q. That is, we want to know how small

$$\left| \alpha - \frac{p}{q} \right| \tag{3.4}$$

can be made by a suitable choice of the rational p/q. Unfortunately, this problem is too easy to be of any interest: as the rationals are dense in $\mathbb{R}$, the difference (3.4), for any α, can be made as small as desired by choosing a large value of q and an appropriate p. Specifically, if we want the difference to be smaller than a positive number ε, we choose $q > 1/2\varepsilon$ and let p be the closest integer to $q\alpha$. Then

$$|q\alpha - p| \le \frac{1}{2} \quad \Rightarrow \quad \left| \alpha - \frac{p}{q} \right| \le \frac{1}{2q} < \varepsilon . \tag{3.5}$$

This observation, though not very interesting in itself, may suggest a more significant approach, namely, to insist that the closeness of approximation should depend on the denominator of the approximating fraction. In other words, we shall be interested in a fairly weak approximation if it is given by a fraction with very small denominator, whereas if the denominator is large we shall expect the approximation to be exceptionally close. One way to achieve this is to try to solve an inequality such as

$$\left| \alpha - \frac{p}{q} \right| < \frac{1}{q^2} \, ,$$

where α is a given real number and we seek rational p/q. In this case, if we are forced to choose a large value of q, we do at least know that the approximation is much closer than we had previously with

$$\left| \alpha - \frac{p}{q} \right| \leq \frac{1}{2q} \, .$$

To obtain worthwhile results we need to note two more points. A single solution of either of the above inequalities is of little importance as it does not give rational numbers *arbitrarily close* to α: what we really want is that there be infinitely many solutions to such an inequality. Secondly, we would like the right-hand side of the inequality to be uniquely determined by the approximating fraction p/q; therefore we shall require that q be the "true" denominator of the fraction, that is, that p and q have no common factor. As a result of these considerations, we introduce the following terminology.

Definition 3.4. *A real number α is said to be* **approximable** *(by rationals) to order s if there exists a constant c such that the inequality*

$$\left| \alpha - \frac{p}{q} \right| < \frac{c}{q^s} \tag{3.6}$$

is satisfied by infinitely rational numbers p/q with p, q relatively prime.

Note.

- In this definition s may be any positive real number. In practice s will frequently, though not always, be an integer.

- The approximations to α given by (3.6) are very close when s is large, less so when s is small. We shall say that α is well approximable by rationals if it is approximable to a high order, and poorly approximable if it is approximable only to a low order.

- It is clear that if α is approximable to order s then it is approximable to any order less than s. So the basic problem in this topic is to find the greatest possible order of approximation for a given real number.

Example. Consider the number

$$\alpha = \sum_{k=0}^{\infty} 10^{-2^k} = 0.1101000100000001000 \cdots ,$$

which we proved irrational in Chapter 1. We observe from the decimal that the rational approximation obtained by taking, say, the first eight decimal places is actually accurate to fifteen decimal places, and therefore is much better than we might expect. If for any m we choose integers

$$q = 10^{2^m} \quad \text{and} \quad p = q \sum_{k=0}^{m} 10^{-2^k} ,$$

then p, q are coprime (look at the last term in the sum) and

$$\left| \alpha - \frac{p}{q} \right| = \frac{1}{10^{2^{m+1}}} + \frac{1}{10^{2^{m+2}}} + \cdots < \frac{2}{10^{2^{m+1}}} = \frac{2}{q^2} .$$

Since m is arbitrary, we can find infinitely many rationals p/q with this property, and so α is approximable to order 2.

There are various slightly different ways to formulate the statement that a number is, or is not, approximable to a given order. A lemma will be useful.

Lemma 3.14. *Given a real number α, let c be a real constant, s and Q positive real numbers. Then the inequalities*

$$\left| \alpha - \frac{p}{q} \right| < \frac{c}{q^s} \quad \text{and} \quad 0 < q < Q$$

are satisfied simultaneously by only a finite number of pairs (p, q) of integers.

Proof. Let (p, q) satisfy the inequalities. Clearly there are only finitely many possible values for q, and since

$$q\alpha - \frac{c}{q^{s-1}} < p < q\alpha + \frac{c}{q^{s-1}} ,$$

each of these yields only finitely many p. This proves the lemma.

Theorem 3.15. Alternative definitions. *Let α be a real number and s a positive real number. Then the following are equivalent:*

- *α is approximable to order s;*
- *there exists a constant c such that the inequality*

$$0 < \left| \alpha - \frac{p}{q} \right| < \frac{c}{q^s} \tag{3.7}$$

  is satisfied by pairs (p, q) with arbitrarily large q;
- *there exists a constant c such that (3.7) is satisfied by infinitely many pairs of integers (p, q).*

Proof. Suppose that α is approximable to order s, so that infinitely many pairs (p, q) of coprime integers satisfy (3.6). By Lemma 3.14, only finitely many of these satisfy $q < Q$, where Q is any given bound. So (3.6) has solutions (p, q) with arbitrarily large q. The equality $|\alpha - p/q| = 0$ holds for only one of these pairs – indeed, for none of them if α is irrational – and so pairs with arbitrarily large q satisfy (3.7). This shows that the first statement implies the second.

It is clear that the second statement implies the third; now we show that the third implies the first. If (mp, mq) is a solution of (3.7), then so is (p, q), for we have

$$0 < \left| \alpha - \frac{p}{q} \right| = \left| \alpha - \frac{mp}{mq} \right| < \frac{c}{(mq)^s} \leq \frac{c}{q^s} .$$

Therefore, if (3.7) has infinitely many solutions but (3.6) has only finitely many, then there would be a pair (p, q) such that (mp, mq) satisfies (3.7) for infinitely many m. Then we should have

$$0 < \left| \alpha - \frac{p}{q} \right| < \frac{c}{m^s q^s}$$

for arbitrarily large m; but this is impossible, as $|\alpha - p/q|$ is a fixed real number, while $c/m^s q^s$ tends to zero as $m \to \infty$. This shows that the third statement implies the first, and completes the proof.

Comment. We need to include the left–hand inequality in (3.7) since we are now speaking not of rational numbers but of pairs of integers, and have dropped the requirement that p, q be coprime. This means that if $\alpha = a/b$ is rational, the inequality

$$\left| \alpha - \frac{p}{q} \right| < \frac{c}{q^s}$$

would be trivially satisfied by the infinitely many pairs $(p, q) = (ma, mb)$.

Theorem 3.16. A non–approximability criterion. *Let α be a real number and t a positive real number. Suppose that there exists a positive constant c such that*

$$\left| \alpha - \frac{p}{q} \right| \geq \frac{c}{q^t}$$

for all rational numbers $p/q \neq \alpha$. Then α is not approximable to any order greater than t.

Proof. If the given assumption holds and α is approximable to order $s > t$ then, using the second part of Theorem 3.15, there is another constant c' such that the inequalities

$$\frac{c}{q^t} \leq \left| \alpha - \frac{p}{q} \right| < \frac{c'}{q^s}$$

are satisfied by certain rationals p/q with arbitrarily large q. But the inequalities yield $q^{s-t} < c'/c$; as the exponent $s - t$ is positive, we obtain a contradiction by choosing q sufficiently large.

We have now seen enough generalities concerning approximability to begin looking at some specific cases. First, the reasoning which led to the inequality (3.5) gives us a result valid for all real numbers.

Theorem 3.17. *Any real number is approximable to order* 1, *at least.*

Proof. Take, say, $c = 1$ and for any q let p be the closest integer to $q\alpha$. Then

$$\left| \alpha - \frac{p}{q} \right| = \frac{|q\alpha - p|}{q} \leq \frac{1}{2q} < \frac{c}{q} \, .$$

If α is irrational, then the left–hand side is non–zero for all q; if α is rational, it is non–zero for all q except multiples of the denominator of α; in either case (3.7), with $t = 1$, is true for infinitely many pairs (p, q), and so α is approximable to order 1.

Next we consider the case $\alpha \in \mathbb{Q}$. Of course such an α can be "approximated" precisely by one rational number, namely, α itself. However, our definition of approximability demands that there be infinitely many distinct rationals close to α. In this sense rational numbers are only very weakly approximable.

Lemma 3.18. *Let α be rational. Then there exists a constant $c > 0$ such that for every rational $p/q \neq \alpha$ we have*

$$\left| \alpha - \frac{p}{q} \right| \geq \frac{c}{q} \, .$$

Proof. Let $\alpha = a/b$. If $p/q \neq a/b$, then $aq - pb$ is a non–zero integer and so

$$\left| \alpha - \frac{p}{q} \right| = \frac{|aq - pb|}{bq} \geq \frac{1}{bq} \, .$$

Taking $c = 1/b$, the lemma is proved.

Comments.

- We say that "there exists a constant c" in order to get used to the style of future proofs. Although in the present case, we have given an explicit formula for c, to do so in more advanced results is inconvenient at best, sometimes difficult, sometimes even impossible.

- We can use this lemma to prove the irrationality of e. For suppose that $e \in \mathbb{Q}$, and for any $n \geq 1$ choose

$$q = n! \, , \quad p = n! \left(1 + \frac{1}{1!} + \frac{1}{2!} + \cdots + \frac{1}{n!} \right) .$$

Then there is a positive constant c such that for all n, we have

$$\frac{c}{q} \leq \left| e - \frac{p}{q} \right| = \frac{1}{(n+1)!} + \frac{1}{(n+2)!} + \cdots < \frac{1}{(n+1)!} \frac{n+1}{n} = \frac{1}{nq} \, ;$$

but this cannot be true when $n > 1/c$. Essentially, this is the same as the proof we gave in Chapter 1.

Theorem 3.19. *A rational number is not approximable to any order $s > 1$.*

Proof. This is an immediate consequence of the previous lemma and the non–approximability criterion, Theorem 3.16. However, we shall repeat the argument for the sake of clarity. Suppose, then, that α is approximable to order $s > 1$. By using the above lemma and one of our equivalent definitions of approximability, we have

$$\frac{c}{q} \le \left| \alpha - \frac{p}{q} \right| < \frac{c'}{q^s}$$

for some positive constants c, c' and for certain p/q with arbitrarily large q. But the inequality implies $q^{s-1} < c'/c$ with $s - 1 > 0$, and for large q this is impossible.

Example. The irrationality of

$$\alpha = \sum_{k=0}^{\infty} 10^{-2^k} = 0.1101000100000001000 \cdots$$

follows immediately, since we showed on page 43 that α is approximable to order 2. Observe that although the decimal expansion of α provides the motivation for the irrationality proof, the actual details are entirely independent of the decimal. Therefore, this proof is quite different from the one we gave in Chapter 1.

Theorem 3.19 is entirely characteristic of rational numbers; irrational numbers, on the other hand, are approximable to order 2 or more. The proof of the next lemma is a beautiful and surprising application of the pigeonhole principle.

Lemma 3.20. *Let $\alpha \in \mathbb{R}$ and let Q be a positive integer. Then there is a rational number p/q in lowest terms, with denominator at most Q, such that*

$$\left| \alpha - \frac{p}{q} \right| < \frac{1}{Qq} \ .$$

Proof. Divide the interval $[0, 1)$ into Q subintervals

$$\left[0, \frac{1}{Q} \right), \ \left[\frac{1}{Q}, \frac{2}{Q} \right), \dots, \left[\frac{Q-1}{Q}, 1 \right),$$

and consider the fractional parts of the $Q+1$ numbers $0, \alpha, 2\alpha, \dots, Q\alpha$. These fractional parts all lie in the interval $[0, 1)$, and so two of them must lie in the same one of the Q subintervals mentioned above. Therefore, there exist integers q_1 and q_2 with $0 \le q_1 < q_2 \le Q$ and

$$\big| (q_2 \alpha - \lfloor q_2 \alpha \rfloor) - (q_1 \alpha - \lfloor q_1 \alpha \rfloor) \big| < \frac{1}{Q} \, .$$

Taking $q = q_2 - q_1$ and $p = \lfloor q_2 \alpha \rfloor - \lfloor q_1 \alpha \rfloor$, we have $0 < q \le Q$ and

$$\big| q\alpha - p \big| < \frac{1}{Q} \, .$$

If we divide out any common factor of p, q the inequality remains true, and the lemma follows.

Theorem 3.21. Approximability of irrational numbers. *Any irrational real number is approximable to order* 2, *at least.*

Proof. Let α be irrational. Take any $Q = Q_1 > 0$ and choose a rational number p_1/q_1 satisfying the lemma. Then

$$\left| \alpha - \frac{p_1}{q_1} \right| < \frac{1}{Q_1 q_1} \le \frac{1}{q_1^2} \, ,$$

and the left–hand side is not zero since $\alpha \notin \mathbb{Q}$. Now take

$$Q = Q_2 > \left| \alpha - \frac{p_1}{q_1} \right|^{-1}$$

and apply the lemma to find a rational number p_2/q_2. Again we have

$$0 < \left| \alpha - \frac{p_2}{q_2} \right| < \frac{1}{q_2^2} \, ;$$

moreover,

$$\left| \alpha - \frac{p_2}{q_2} \right| < \frac{1}{Q_2} < \left| \alpha - \frac{p_1}{q_1} \right|$$

and so p_2/q_2 is not equal to p_1/q_1. Continuing in this way we find infinitely many solutions of the inequality (3.7) with $s = 2$.

Just as rational numbers are distinguished from irrationals in being approximable to no greater order than 1, so, as it turns out, algebraic numbers are distinguished from (some) transcendentals in being approximable to no greater order than 2. The following result was proved by K.F. Roth in 1955, and earned him the 1958 Fields medal.

Theorem 3.22. Roth's Theorem [56]. *A real algebraic number is not approximable to any order greater than* 2.

Roth's Theorem was the culmination of a series of results on approximability properties of algebraic numbers, of which the most important are the following. If α is an algebraic number of degree n, then

- Liouville [39] proved in 1851 that α is not approximable to order greater than n;

- in 1908–9 Thue [63], [64] improved Liouville's result by showing that α is not approximable to order greater than $\frac{1}{2}n + 1$;

- Siegel [59] showed in 1921 that α is not approximable to order greater than $2\sqrt{n}$.

Since we have already shown that every irrational number is approximable to order 2, Roth's Theorem is the best possible result of its type. We shall prove Liouville's result, which suffices to establish the transcendence of certain interesting real numbers, but shall pass over with brief comments the results of Thue, Siegel and Roth.

K.F. Roth
(1925–2015)

Theorem 3.23. (Liouville, 1851). *If α is a real algebraic number of degree n, then α is not approximable to any order exceeding n.*

Proof. If α is rational, then α has degree 1, and we have already proved the result (Theorem 3.19). So let α be an irrational algebraic number with minimal polynomial f and degree $n \geq 2$. For any real x, the Mean Value Theorem states that

$$\frac{f(x) - f(\alpha)}{x - \alpha} = f'(\gamma)$$

for some γ between α and x. Taking $x = p/q$, rearranging and recalling that $f(\alpha) = 0$ by definition, we have

$$f\left(\frac{p}{q}\right) = -f'(\gamma)\left(\alpha - \frac{p}{q}\right) . \tag{3.8}$$

We wish to obtain a lower estimate for $q^n |\alpha - p/q|$, so that we can establish Liouville's result by appealing to the non–approximability criterion, Theorem 3.16. Note, however, that the constant c in that theorem is independent of p/q; here γ does depend on p/q and therefore must not appear in our estimate. First, suppose that $|\alpha - p/q| \leq 1$. Then $\alpha - 1 \leq \gamma \leq \alpha + 1$. Since f' is continuous on $\mathbb{R}$ it is bounded on any finite interval and we have, say, $|f'(\gamma)| < \mu$. From the equality (3.8) we obtain

$$\left|f\left(\frac{p}{q}\right)\right| \leq \mu \left|\alpha - \frac{p}{q}\right| .$$

Now let b be a common denominator for the coefficients of f. Then $bf(x)$ is a polynomial with integer coefficients and so $bq^n f(p/q)$ is an integer, which, moreover, is not zero since f is irreducible and has no rational roots. Therefore

$$\left| \alpha - \frac{p}{q} \right| \geq \frac{1}{\mu} \left| f\left(\frac{p}{q} \right) \right| \geq \frac{1}{b\mu q^n} \,,$$

provided that $|\alpha - p/q| \leq 1$. On the other hand, if $|\alpha - p/q| > 1$ then obviously

$$\left| \alpha - \frac{p}{q} \right| \geq \frac{1}{q^n} \,.$$

So take $c = \min(1, 1/b\mu)$. Then since c does not depend on p/q, we have for every rational number

$$\left| \alpha - \frac{p}{q} \right| \geq \frac{c}{q^n} \,, \tag{3.9}$$

and by the non–approximability criterion, this shows that α is not approximable to order greater than n.

At last we can give a specific example of a transcendental number.

Theorem 3.24. Liouville's number. *The number*

$$\lambda = \sum_{k=1}^{\infty} 10^{-k!}$$

is transcendental.

Proof. Suppose, on the contrary, that λ is algebraic of degree n. For any integer $m \geq n$ define

$$q = 10^{m!} \quad \text{and} \quad p = q \sum_{k=1}^{m} 10^{-k!} \,.$$

Then p and q are coprime integers and we have

$$\left| \lambda - \frac{p}{q} \right| = \frac{1}{10^{(m+1)!}} + \frac{1}{10^{(m+2)!}} + \cdots < \frac{2}{10^{(m+1)!}} = \frac{2}{q^{m+1}} \leq \frac{2}{q^{n+1}} \,.$$

Since this holds for all $m \geq n$, there are infinitely many rational p/q in lowest terms satisfying

$$\left| \lambda - \frac{p}{q} \right| < \frac{c}{q^s}$$

with $c = 2$ and $s = n + 1$, and so λ is approximable to order $n + 1$. But this contradicts Liouville's Theorem, and it follows that λ is transcendental.

Definition 3.5. *A real number which is approximable to arbitrarily high order is called a* **Liouville number.**

It follows immediately from Theorem 3.23 that every Liouville number is transcendental. The converse, however, is not true. Consider once again

$$\alpha = \sum_{k=0}^{\infty} 10^{-2^k} = 0.1101000100000001000\cdots.$$

By truncating the decimal after each 1 (the same idea, in fact, as we used for λ) we showed that α is approximable to order 2. On the other hand, it is difficult to think of any way of obtaining better rational approximations than these; this suggests that α is not approximable to any order greater than 2, and so is not a Liouville number. However, α is transcendental, as we shall show in Chapter 6. As an example of a transcendental number which is (demonstrably!) not a Liouville number, we need only consider π. Kurt Mahler [42] proved in 1953 that

$$\left| \pi - \frac{p}{q} \right| > \frac{1}{q^{42}}$$

for all rational numbers p/q with $q \geq 2$; this result has been improved by various authors, V.Kh. Salikhov [57, 58] showing in 2008 that there is a constant q_0 such that

$$\left| \pi - \frac{p}{q} \right| > \frac{1}{q^{7.61}}$$

for all p/q with $q \geq q_0$. Therefore, π is not approximable to order greater than 7.61. Either of these results shows that π is not a Liouville number. However, π is transcendental: a proof of this will be given in Chapter 5 by extending Hermite's methods from Chapter 2.

In 2020 Zeilberger and Zudilin improved upon some of the details in Salikhov's argument, thereby marginally reducing the exponent 7.61 to 7.11. At the time of writing, this is the record in what the authors [71] refer to as a "competitive sport".

The exponential constant e is another well-known number which is transcendental (see Chapter 5) but not Liouville (exercise 4.20). Some very different examples were obtained by Mahler [41] in 1937: he showed that the members of a class of real numbers represented by decimal expansions such as

$$0.12345678910111213141516\cdots \quad \text{and} \quad 0.149162536496481100121144\cdots$$

– the former is the Champernowne constant mentioned in Chapter 1 – are transcendental but are not Liouville numbers. In Chapter 4 we shall use continued fractions to give a complete proof of the existence of non–Liouville transcendental numbers.

Note the importance, in Liouville's proof, of ensuring that the constant c in the inequality (3.9) is independent of p/q. This is necessary since the inequality in Theorem 3.16 must hold for a *fixed* c and *all* rationals p/q except α. If c is allowed to depend on p/q, the inequality can be satisfied trivially, regardless of the approximability properties of α. In Liouville's proof, γ lies between α and p/q. Therefore, although no simple functional relationship has been given, γ does depend on p/q, and so c must not depend directly on γ. To overcome this problem we replace $|f'(\gamma)|$ by μ, the maximum absolute value of f' over the interval $[\alpha - 1, \alpha + 1]$, which is independent of p/q.

On the other hand, it *is* permissible for c to depend on α and on s. Considering Liouville's proof again, the algebraic number α completely determines the polynomial f and hence the common denominator b; the derivative f' therefore is also uniquely determined by α, and so is μ. Hence $c = \min(1, 1/b\mu)$ defines c as a positive real number which depends on α but not on any particular rational p/q.

Similar considerations apply if we wish to show from the definition that a real number α is approximable to order s: the constant c in the inequality (3.6) appearing in the definition must be the same for all solutions p/q. This is obviously so in, for example, the proof that every irrational is approximable to order 2, where we just took $c = 1$.

Liouville's Theorem can be generalised in various ways. First, we rephrase the result (3.9) that we obtained on page 50, near the end of Liouville's proof.

Theorem 3.25. *Let α be an algebraic number of degree n. Then there is a constant c, depending only on α, such that for all integers p and all positive integers q we have*

$$\left| q\alpha - p \right| \geq \frac{c}{q^{n-1}} \; ,$$

provided that $q\alpha - p \neq 0$.

This can be regarded as a result about the size of $g(\alpha)$, where $g(z) = qz - p$ is a linear polynomial with integer coefficients. We may generalise the result by removing the restriction that g be linear. As on page 41, let the **height** of any polynomial $g \in \mathbb{Z}[x]$ be the maximum of the absolute values of the coefficients of g. That is, if $g(z) = a_m z^m + a_{m-1} z^{m-1} + \cdots + a_1 z + a_0$ then

$$H(g) = \max_{0 \leq k \leq m} |a_k| \; .$$

We have the following result.

Theorem 3.26. *Let α be an algebraic number of degree n. Then there is a positive constant c, depending only on α, such that for all polynomials g of degree m, with integer coefficients, we have*

$$\left| g(\alpha) \right| \geq \frac{c^m}{H(g)^{n-1}} \; ,$$

provided that $g(\alpha) \neq 0$.

Proof. Let the conjugates of α be $\alpha_1, \alpha_2, \ldots, \alpha_n$ with $\alpha = \alpha_1$, and let a be the maximum of the absolute values of these conjugates,

$$a = \max_{1 \leq k \leq n} |\alpha_k| \, .$$

From the definition of $H(g)$, we have

$$
\begin{aligned}
|g(\alpha_k)| &\leq |g_m||\alpha_k|^m + |g_{m-1}||\alpha_k|^{m-1} + \cdots + |g_1||\alpha_k| + |g_0| \\
&\leq H(g)(a^m + a^{m-1} + \cdots + a + 1) \\
&\leq H(g)(a+1)^m
\end{aligned}
$$

for each k. Let d be a common denominator for $\alpha_1, \alpha_2, \ldots, \alpha_n$, that is, a positive rational integer such that $d\alpha_k$ is an algebraic integer for every k. Then

$$d^m g(\alpha_k) = g_m(d\alpha_k)^m + dg_{m-1}(d\alpha_k)^{m-1} + \cdots + d^{m-1}g_1(d\alpha_k) + d^m g_0$$

is a non–zero algebraic integer for every k. It follows that

$$N = \prod_{k=1}^{n} d^m g(\alpha_k) \tag{3.10}$$

is a non–zero algebraic integer. By using properties of symmetric polynomials – see Chapter 5 – it is possible to show that the product on the right–hand side of (3.10) is rational; consequently N is a rational integer and has absolute value at least 1. Splitting the first term off from the product and rearranging the powers of d, we have

$$d^{mn} |g(\alpha)| \prod_{k=2}^{n} |g(\alpha_k)| \geq 1 \, .$$

Now for $k = 2, 3, \ldots, n$, the estimate for $|g(\alpha_k)|$ obtained above yields

$$d^{mn} |g(\alpha)| H(g)^{n-1} (a+1)^{m(n-1)} \geq 1 \, ;$$

choose

$$c = \frac{1}{d^n (a+1)^{n-1}} \, .$$

Then c depends on n, a and d, all of which are uniquely determined by α. Therefore, c depends on α alone, and since

$$|g(\alpha)| \geq \frac{c^m}{H(g)^{n-1}}$$

the result is proved.

Another way to generalise Liouville's Theorem is to regard it as providing information about the approximation of a fixed algebraic number α by algebraic numbers of degree 1. The following result, which is proved in Lipman [40], concerns the approximation of a given algebraic number by other algebraic numbers.

Theorem 3.27. *Let α be an algebraic number of degree n. Then there is a positive constant c, depending only on α, such that for all algebraic numbers ξ of degree m we have*

$$|\alpha - \xi| \geq \frac{c^m}{H(\xi)^n} \,,$$

provided that $\xi \neq \alpha$. Here the algebraic number ξ satisfies a unique polynomial g of degree m having relatively prime integer coefficients, and the height $H(\xi)$ is defined to be the height of g, in the sense given on page 41.

We complete this section with a brief description of the methods used by Thue and Roth to improve Liouville's Theorem. Thue's result that a real algebraic number of degree n is not approximable to order exceeding $\frac{1}{2}n + 1$ is proved by contradiction. Assume that α has degree n and is approximable to order $s > \frac{1}{2}n + 1$. That is, there is a constant c such that the inequality

$$0 < \left| \alpha - \frac{p}{q} \right| < \frac{c}{q^s} \tag{3.11}$$

has infinitely many solutions. Choose a solution p_1/q_1 with q_1 very large. Let m be a large integer, and choose another solution p_2/q_2 of (3.11), taking q_2 so large by comparison with q_1 that $q_2 \approx q_1^m$. Thue then shows that it is possible to find a polynomial $g(x, y)$ in two variables, with integral coefficients, having the following properties: (i) $g(\alpha, \alpha) = 0$; (ii) the partial derivatives $\partial^k g / \partial x^k$, evaluated at (α, α), are zero for many values of k; (iii) the coefficients of g are "not too large"; (iv) $g(p_1/q_1, p_2/q_2)$ is not zero. In fact, if g_1 and g_2 are polynomials then

$$g(x, y) = (y - \alpha)g_1(x, y) + (x - \alpha)^m g_2(x, \alpha)$$

clearly satisfies condition (i), and because of the factor $(x - \alpha)^m$, condition (ii) is a restriction only on g_1, and not on g_2. Thue proved the existence of g_1 so that (ii) is satisfied, and of g_2 so that (iii) and (iv) are then also satisfied – this is the difficult bit. Once g has been found, the proof proceeds in a manner similar to that of Liouville.

Siegel showed that an algebraic number is not approximable to order greater than $2n^{1/2}$. Assuming that inequality (3.11) has infinitely many solutions for some $s > 2n^{1/2}$, Siegel chose two solutions with large denominators (as Thue had done), and obtained a contradiction by refining certain details in Thue's argument. The belief arose that for $s > c_1 n^{1/k}$, one should be able to choose k solutions of (3.11) and then use polynomials in k variables to obtain a contradiction. If this could be done for arbitrarily large k, the greatest possible order of approximation s would simply be a constant, irrespective of the degree of α. The construction of the required polynomial turned out to be very difficult but was eventually accomplished in 1955 by K.F. Roth.

Theorem 3.22. Roth's Theorem [56]. *A real algebraic number is not approximable to any order greater than 2.*

3.4 IRRATIONALITY OF $\zeta(3)$: A SKETCH

We conclude this chapter with a very brief summary of Apéry's notoriously complex irrationality proof for $\zeta(3)$. Our only aim is to show how the argument is based fundamentally on the approximation ideas introduced in the previous section: specifically, Apéry showed that $\zeta(3)$ is approximable to order (just slightly) greater than 1, and therefore cannot be rational. The reader should not be deluded into believing that the arguments we have omitted are easy! – they most assuredly are not. More details (as well as an engaging account of the circumstances surrounding Apéry's announcement of his result) may be found in [66]. Another, and possibly simpler, irrationality proof for $\zeta(3)$ was given by Beukers [14]. Although superficially Beukers' approach appears quite different from Apéry's, the author acknowledges a close connection between the two.

So, we begin by recalling the definition

$$\zeta(3) = \sum_{n=1}^{\infty} \frac{1}{n^3} = 1 + \frac{1}{2^3} + \frac{1}{3^3} + \frac{1}{4^3} + \cdots .$$

By intricate but essentially straightforward algebra we may obtain an alternative summation formula

$$\zeta(3) = \frac{5}{2} \sum_{n=1}^{\infty} \frac{(-1)^{n-1}}{n^3 \binom{2n}{n}} = \frac{5}{2} \left(\frac{1}{1^3 \times 2} - \frac{1}{2^3 \times 6} + \frac{1}{3^3 \times 20} - \frac{1}{4^3 \times 70} + \cdots \right) .$$

The heart of Apéry's argument consists of defining two "mysterious" sequences a_n and b_n with the property that the quotients a_n/b_n form a sequence of very good rational approximations to $\zeta(3)$. Set

$$a_0 = 0 , \; a_1 = 6 , \; a_n = \frac{34n^3 - 51n^2 + 27n - 5}{n^3} a_{n-1} - \frac{(n-1)^3}{n^3} a_{n-2}$$

for $n \geq 2$, and

$$b_0 = 1 , \; b_1 = 5 , \; b_n = \frac{34n^3 - 51n^2 + 27n - 5}{n^3} b_{n-1} - \frac{(n-1)^3}{n^3} b_{n-2}$$

for $n \geq 2$. One observes that a_n and b_n satisfy the same recurrence, and differ only in their respective initial conditions. Amazingly, despite the fractional coefficients in its recurrence, it can be shown that b_n is always an integer! This is not the case for a_n; however, it turns out that a_n is a rational number whose denominator is a factor of $2L_n^3$, where L_n is the least common multiple of the integers $1, 2, \ldots, n$. Apéry also proved that

$$\lim_{n \to \infty} \frac{a_n}{b_n} = \zeta(3) .$$

The above results about a_n and b_n are the most difficult aspects of Apéry's argument. It is then relatively straightforward to prove that

$$\zeta(3) - \frac{a_n}{b_n} = \sum_{k=n+1}^{\infty} \frac{6}{k^3 b_k b_{k-1}} \, ,$$

from which it follows that

$$0 < \zeta(3) - \frac{a_n}{b_n} < \frac{c_1}{b_n^2}$$

for a certain constant c_1. This *does not* show that $\zeta(3)$ is approximable to order 2, because a_n is not an integer. But if we define

$$p_n = 2L_n^3 a_n \, , \quad q_n = 2L_n^3 b_n \, ,$$

then p_n and q_n are integers and we have

$$0 < \left| \zeta(3) - \frac{p_n}{q_n} \right| < \frac{c_1}{b_n^2} \, .$$

Our aim is to find a constant $s > 1$ such that

$$\frac{1}{b_n^2} < \frac{1}{q_n^s} \tag{3.12}$$

for infinitely many n. To achieve this we need to know something about the size of b_n; and also of L_n, since that appears in the definition of q_n. It can be shown that if

$$\lambda = 17 + 12\sqrt{2}$$

(not actually as mysterious as it seems!) then

$$\frac{b_n}{\lambda^n / n^{3/2}}$$

approaches a finite non–zero limit as $n \to \infty$; hence there exist constants $c_2, c_3 > 0$ such that

$$\frac{c_2 \lambda^n}{n^{3/2}} < b_n < \frac{c_3 \lambda^n}{n^{3/2}} \, .$$

The *Prime Number Theorem*, which gives an estimate for the number of primes not exceeding n, can be employed to show that

$$L_n < 3^n$$

for all sufficiently large n. (See appendix 3.3 for some details and further references.) A little easy algebra then shows that the inequality (3.12) is true for all sufficiently large n satisfying

$$\left(\frac{\lambda^2}{(27\lambda)^s} \right)^n > \frac{4c_3^2}{c_2^2} n^{3 - \frac{3}{2}s} \, .$$

Now the left–hand side is an exponential, the right–hand side is a power function; so the left–hand side is certainly the greater (for sufficiently large n), provided that it is an *increasing* exponential, that is, $\lambda^2 > (27\lambda)^s$. This holds for some $s > 1$ if and only if $\lambda > 27$; referring back to the value of λ stated above, it is very easy to see that this is true, and the proof is essentially complete. Specifically, any s greater than 1 and less than

$$\frac{2\log\lambda}{\log(27\lambda)} = 1.033667\cdots$$

will do what we want; we have shown, therefore, that

$$0 < \left|\zeta(3) - \frac{p}{q}\right| < \frac{c_1}{q^{1.03}}$$

has infinitely many solutions, so $\zeta(3)$ is approximable to order 1.03 and cannot be rational.

EXERCISES

3.1 Find the minimal polynomial of $\alpha = \sqrt{2} + \sqrt[3]{3}$.

3.2 Find the minimal polynomial of $\alpha = \cos\frac{1}{9}\pi$. What are the conjugates of $\cos\frac{1}{9}\pi$?

3.3 If $\zeta = e^{2\pi i/5}$ is a fifth root of unity, find the minimal polynomials of $\alpha = \zeta + \zeta^4$ and $\beta = \zeta + \zeta^2$.

3.4 Prove that $a_n z^n + a_{n-1} z^{n-1} + \cdots + a_1 z + a_0$ is irreducible if and only if $a_0 z^n + a_1 z^{n-1} + \cdots + a_{n-1} z + a_n$ is irreducible.

3.5 Let $n > 2$ be an integer.

   (a) Show that $\tan(\pi/n)$ is algebraic.

   (b) Find the minimal polynomial of $\tan(\pi/p)$ if p is prime and $p > 2$.

   (c) Show that if n is composite, then the degree of $\tan(\pi/n)$ is strictly less than $n - 1$. Find the minimal polynomial of $\tan(\pi/10)$.

3.6 Suppose that α is a root of a polynomial $a_n z^n + a_{n-1} z^{n-1} + \cdots + a_0$ having integral coefficients. Prove that the denominator of α is a factor of a_n.

3.7 Let 3θ be the smallest angle of a 3–4–5 triangle. Show that $\alpha = \cos\theta$ is an algebraic number; find its degree, its denominator and its conjugates.

3.8 Show that if a, b are positive rationals, $a \neq b$ and $a^{1/n} - b^{1/n}$ is rational, then $a^{1/n}$ and $b^{1/n}$ are both rational.

   This was set as a "challenge problem" in exercise 1.4 – congratulations to any reader who solved it! The problem should be somewhat more approachable using results we have seen in Chapter 3.

3.9 Prove that 1, $\sqrt[3]{2}$, $\sqrt[3]{4}$ are linearly independent over $\mathbb{Q}$. This problem appeared as exercise 1.6: it should be quite easy using methods of the present chapter.

3.10 (a) Show that two conjugate algebraic numbers can never differ by a rational number. (Except for zero of course!)

(b) Let α be algebraic and not zero. Prove that if r is rational and $r\alpha$ is a conjugate of α, then $r = \pm 1$; furthermore, if α has odd degree, then $r = 1$ only.

3.11 Let the polynomial $f(z)$ be irreducible over $\mathbb{Q}$. Show that $f(z)$ has no repeated factors over $\mathbb{C}$.

Comment. The terminology in use is that over $\mathbb{Q}$, every irreducible polynomial is **separable**. This result does actually depend on specific properties of $\mathbb{Q}$, and there exist cases in which an analogous result is not true. See [61], Chapter 8.

3.12 Show that for any integer $n \geq 2$ and any prime number p, the polynomial $f(z) = z^n + z^{n-1} + p$ is irreducible.

3.13 In this exercise we investigate the irreducibility of $f(z) = z^4 + 1$.

(a) Suppose that p is a prime and that there exists an integer a such that $a^2 \equiv 2 \pmod{p}$. Factorise $f_p(z)$ as a product of two quadratics over $\mathbb{Z}_p$.

(b) Suppose that p is prime and there exists a with $a^2 \equiv -2 \pmod{p}$. Factorise $f_p(z)$ over $\mathbb{Z}_p$.

(c) It can be shown (see, for example, Hardy and Wright [29], section 6.7) that if p is prime and neither of the above conditions holds, then there exists a such that $a^2 \equiv -1 \pmod{p}$. Use this fact to complete the proof that if p is any prime, then f_p is reducible over $\mathbb{Z}_p$.

(d) Find a composite m such that f_m is irreducible over $\mathbb{Z}_m$, and deduce that f is irreducible over $\mathbb{Z}$.

(e) Alternatively, use Eisenstein's criterion to show that f is irreducible over $\mathbb{Z}$.

3.14 (a) Prove the following extension of Eisenstein's Lemma. Let

$$f(z) = a_n z^n + a_{n-1} z^{n-1} + \cdots + a_1 z + a_0$$

be a polynomial with integral coefficients and let m be an integer, $1 \leq m \leq n$. Suppose there exists a prime p with $p \mid a_0, a_1, \ldots, a_{m-1}$ and $p \nmid a_m$ and $p^2 \nmid a_0$. Then f is not the product of two rational polynomials of degree less than m.

(b) Deduce that $f(z) = z^4 + z^3 + 2z^2 + 6z + 2$ is irreducible.

3.15 Prove *Westlund's criteria* for irreducibility [68]. Let $a_1, a_2, \ldots, a_n$ be distinct integers. Then

- $f(z) = (z - a_1)(z - a_2) \cdots (z - a_n) - 1$ is irreducible;
- $g(z) = (z - a_1)(z - a_2) \cdots (z - a_n) + 1$ is irreducible unless it is a perfect square.

3.16 Prove that if at least one of the numbers α and β is transcendental, then at least one of the numbers $\alpha + \beta$ and $\alpha\beta$ is transcendental.

3.17 Prove that if α is not approximable to order s, then there exists a positive real constant c such that the inequalities

$$0 < \left| \alpha - \frac{p}{q} \right| < \frac{c}{q^s}$$

have no rational solution p/q.

3.18 Show that if α is approximable to order s, then α^2 is approximable to order $s/2$.

3.19 Show that if a and s are positive integers with $a \geq 2$ then

$$\alpha = \sum_{k=1}^{\infty} \frac{1}{a^{s^k}}$$

is approximable to order s. Deduce from Roth's Theorem that if $s \geq 3$ then α is transcendental.

3.20 *A simultaneous approximation problem.* Prove that there is a positive integer $k \leq 100$ such that $k\sqrt{2}$ and $k\sqrt{3}$ are both integers to within 0.1.

3.21 Let a be a natural number with $a \geq 2$, and let $\{ b_k \}_{k=1}^{\infty}$ be a sequence of integers satisfying

$$0 < b_1 < b_2 < b_3 < \cdots .$$

Show that if b_{k+1}/b_k is unbounded then

$$\alpha = \sum_{k=1}^{\infty} \frac{1}{a^{b_k}}$$

is a Liouville number (and hence is transcendental).

3.22 Prove that any real number is the sum of two Liouville numbers.

3.23 *Ruler–and–compass constructions.* Two of the great unsolved problems of geometry in Ancient Greece were known as "duplicating the cube" and "trisecting the angle". In each case the constructions were to be accomplished without the use of any instruments other than a pair of compasses and an unmarked ruler. In the nineteenth century these problems were approached from an algebraic point of view, and it was shown (see, for example, [61]) that, under the stated restrictions, line segments with lengths in the ratio α can be constructed only if α is an algebraic number whose degree is a power of 2.

(a) The problem of *duplicating the cube* is: given a cube, construct a cube of twice its volume. Use the result mentioned above to show that this construction is impossible (by means of ruler and compass).

(b) The problem of *trisecting the angle* is: given any angle, divide it into three equal angles. Explain why constructing an angle of size θ is equivalent to constructing line segments in the ratio $\cos \theta$. Then suppose that $\cos \theta$ is rational, and show that θ can be trisected by ruler and compasses only if the polynomial $f(z) = 4z^3 - 3z - \cos \theta$ is reducible. Finally, find an example of an angle which cannot be trisected with ruler and compasses.

3.24 Find a set of points $\{P_1, P_2, \dots\}$ in the real plane, as few as possible, such that every point in the plane is an irrational distance from at least one P_k. (After [23].)

3.25 Let a, b and c be positive integers, and let α be the real cube root of a/b. By considering rational approximations to α, prove that the equation

$$ax^3 - by^3 = c$$

cannot have infinitely many solutions in integers x, y.

3.26 Let p be prime, $p \geq 3$. Show that if a p-gon has all its angles equal and all its sides of integer length, then it is regular.

3.27 Recall that the Champernowne constant is

$$\xi = 0.12345678910111213 \cdots .$$

The following working shows how to obtain one very good rational approximation to ξ. Use similar ideas to find infinitely many very good rational approximations. What can you say about the transcendence, or, if algebraic, the degree of ξ by appealing to (a) Liouville's Theorem; (b) Siegel's Theorem; (c) Roth's Theorem?

We shall take advantage of the decimal expansion of ξ and related numbers to find a very simple way of multiplying by 99. First, we write

$$10^9 \xi = 123456789.101112 \cdots 99100101 \cdots = p_1 + \xi_1 + r_1 ,$$

where

$$p_1 = 123456789 , \quad \xi_1 = 0.\overbrace{1011\cdots99}^{90 \text{ pairs}}$$

and

$$r_1 = 0.\overbrace{00\cdots00}^{180 \text{ digits}} 100101 \cdots < 10^{-180} .$$

Next we have

$$99\xi_1 = 100\xi_1 - \xi_1$$

$$= 10.\overbrace{1112\cdots9899}^{89 \text{ pairs}} -0.\overbrace{1011\cdots9798}^{89 \text{ pairs}} -0.\overbrace{00\cdots00}^{89 \text{ pairs}} 99$$

$$= p_2 + \xi_2 - r_2 ,$$

where $p_2 = 10$ and

$$\xi_2 = 0.\overbrace{1112\cdots9899}^{89 \text{ pairs}} -0.\overbrace{1011\cdots9798}^{89 \text{ pairs}} = 0.\overbrace{0101\cdots0101}^{89 \text{ pairs}}$$

and $r_2 < 10^{-178}$. Using the same idea again,

$$99\xi_2 = 100\xi_2 - \xi_2 = 1.\overbrace{0101\cdots01}^{88 \text{ pairs}} -0.\overbrace{0101\cdots01}^{89 \text{ pairs}}$$

$$= 1 - r_3$$

with $r_3 = 10^{-178}$. Putting everything back together yields

$$99^2 \times 10^9\xi - 99^2 p_1 - 99p_2 - 1 = 99^2 r_1 - 99r_2 - r_3 ;$$

hence

$$|q\xi - p| \le 99^2 r_1 + 99r_2 + r_3 < \frac{3}{10^{176}} ,$$

where $q = 99^2 \times 10^9$ and p is an integer that we do not need to calculate. It is easy to see that $q < 10^{13}$, which gives

$$\left|\xi - \frac{p}{q}\right| < \frac{3}{q^{1+176/13}} = \frac{3}{q^{14.53\cdots}} .$$

3.28 Prove that

$$\alpha = \left(1 + \frac{1}{2^1}\right)\left(1 + \frac{1}{2^2}\right)\left(1 + \frac{1}{2^6}\right)\cdots = \prod_{k=1}^{\infty}\left(1 + \frac{1}{2^{k!}}\right)$$

is a Liouville number. (You may assume that the infinite product converges.) At some stage the inequality

$$\prod_{k=m+1}^{\infty}\left(1 + \frac{1}{2^{k!}}\right) \le 1 + \frac{2}{2^{(m+1)!}}$$

may be useful (but ideally you should prove this rather than just taking it as given).

APPENDIX 1: COUNTABLE AND UNCOUNTABLE SETS

Definition 3.6. *A set S is said to be* **countable** *if there exists a one–to–one function from S to $\mathbb{N}$, the set of natural numbers, and* **uncountable** *if it is not countable.*

Lemma 3.28. *Any finite set is countable; the set of integers is countable.*

Lemma 3.29. *Let f be a function from A to B and g a function from B to C. If f and g are both one–to–one, so is the composite function $g \circ f$. If $g \circ f$ is one–to–one, then so is f.*

Corollary 3.30. *If T is countable and there is a one–to–one function from S to T, then S is countable.*

Exercise. Give an example to show that if $g \circ f$ is one–to–one, g need not be one–to–one.

Theorem 3.31. *If S and T are countable, then so are $S \times T$ and $S \cup T$. If S is countable and $R \subseteq S$, then R is countable.*

Proof. Let $f : S \to \mathbb{N}$ and $g : T \to \mathbb{N}$ be one–to–one functions. Then

$$h : S \times T \to \mathbb{N}, \quad h(s,t) = 2^{f(s)}3^{g(t)}$$

and

$$k : S \cup T \to \mathbb{N} \times \mathbb{N}, \quad k(x) = \begin{cases} (0, f(x)) & \text{if } x \in S \\ (1, g(x)) & \text{if } x \notin S \end{cases}$$

are one–to–one. If $R \subseteq S$, then the restriction $f \mid_R$ of f to R is one–to–one. The above constructions can easily be generalised to prove the following.

Theorem 3.32. *Suppose that the sets $S_0, S_1, S_2, \ldots$ are countable. Then*

- *$S_0 \times S_1 \times \cdots \times S_n$ is countable for each natural number n;*
- *$S_0 \cup S_1 \cup S_2 \cup \cdots$ is countable.*

Exercise. Explain why we need to restrict the Cartesian product in this result to finitely many terms, but do not need to restrict the union in the same way. The second of the above properties can also be extended a little further.

Theorem 3.33. *Let T be countable and suppose that for every $t \in T$, the set S_t is countable. Then*

$$S = \bigcup_{t \in T} S_t$$

is a countable set.

Proof. Let $g : T \to \mathbb{N}$, and $f_t : S_t \to \mathbb{N}$ for each $t \in T$, be one–to–one functions. For each $x \in S$ let $t = t(x)$ be the element of T such that $g(t)$ is minimal subject to the condition $x \in S_t$. Then

$$h(x) = \big(g(t(x)), f_{t(x)}(x) \big)$$

defines a one–to–one function h from S to $\mathbb{N} \times \mathbb{N}$, and so S is countable.

Theorem 3.34. *If S is countable, then the collection of all finite sequences of elements of S is a countable set.*

Proof. If we write ε for the empty sequence, then the collection we are considering is

$$\{\varepsilon\} \cup S \cup S^2 \cup S^3 \cup \cdots ,$$

and the theorem follows from earlier results.

Theorem 3.35. (Cantor, 1874). *The set of real numbers is uncountable.*

APPENDIX 2: THE MEAN VALUE THEOREM

Theorem 3.36. The Mean Value Theorem of differential calculus. *Let f be a real–valued function defined and continuous on the interval $[a, b] \subseteq \mathbb{R}$, and differentiable on the interior of this interval. Then there is a real number $c \in (a, b)$ such that*

$$\frac{f(b) - f(a)}{b - a} = f'(c) .$$

APPENDIX 3: THE PRIME NUMBER THEOREM

For any real number x, we denote by $\pi(x)$ the number of primes less than or equal to x. For example, $\pi(1) = 0$ and $\pi(10) = 4$ and $\pi(100) = \pi(100.5) = 25$. The symbol π is customary and is used because it is the Greek equivalent of "p", the first letter of the word "prime" – it has, of course, nothing to do with the trigonometric constant π. The following result, which gives an estimate for $\pi(x)$ in terms of elementary functions, was proved independently and more or less simultaneously in 1896 by Hadamard [27] and de la Vallée Poussin [65]. A proof is given by Hardy and Wright [29].

Theorem 3.37. The Prime Number Theorem. *The number of primes not exceeding x satisfies*

$$\frac{\pi(x)}{x / \log x} \to 1 \quad as\ x \to \infty ,$$

where $\log$ denotes the natural (base e) logarithm.

We may use the Prime Number Theorem to estimate the least common multiple of the first n positive integers. Given a prime p and a positive integer n, there is a unique non–negative integer α such that $p^\alpha \leq n < p^{\alpha+1}$; the least common multiple will be the product of all these powers p^α. Therefore

$$L_n = \text{lcm}(1, 2, \ldots, n) = \prod_{\substack{p \leq n \\ p \text{ prime}}} p^\alpha \leq \prod_{\substack{p \leq n \\ p \text{ prime}}} n = n^{\pi(n)} ,$$

the last equality being true because the product consists of $\pi(n)$ equal factors. Let c be a constant greater than e. It follows from the Prime Number Theorem that if n is sufficiently large then

$$\frac{\pi(n)}{n/\log n} < \log c$$

and hence

$$L_n \leq n^{\pi(n)} = e^{\pi(n)\log n} < c^n .$$

In section 3.4 we chose $c = 3$ to keep things simple. We could have taken a slightly smaller value, which would have resulted in a slightly larger value for s; however, this would have made no significant difference to the result.

Comment. In fact, it can be shown [55] that the maximum value of $(L_n)^{1/n}$ occurs when $n = 113$. Therefore, if we take

$$c = \left(\text{lcm}(1, 2, \ldots, 113)\right)^{1/113} = 2.8258821394 \cdots ,$$

then we have $L_n \leq c^n$ for all n, and not just for all sufficiently large n.

CHAPTER 4

Continued Fractions

> *Come back tomorrow night*
> *... we're gonna do fractions!*
>
> Tom Lehrer

I N CHAPTER 3 we were looking for rational approximations to various real numbers as a means of proving irrationality or transcendence. There is in fact a standard procedure for obtaining *all* of the "best" rational approximations (in a sense to be explained later) to a given real number by the use of *continued fractions*.

We shall start with the Euclidean algorithm for computing the greatest common divisor of two (positive) integers. For example, beginning with 95 and 37, we have

$$95 = 2 \times 37 + 21 \ , \quad 37 = 1 \times 21 + 16 \ , \quad 21 = 1 \times 16 + 5 \ , \quad 16 = 3 \times 5 + 1 \ ,$$

which shows that the greatest common divisor of 95 and 37 is 1. Rewriting these equalities in an obvious way gives

$$\frac{95}{37} = 2 + \frac{21}{37} \ , \quad \frac{37}{21} = 1 + \frac{16}{21} \ , \quad \frac{21}{16} = 1 + \frac{5}{16} \ , \quad \frac{16}{5} = 3 + \frac{1}{5} \ ,$$

and if we combine all these expressions we obtain

$$\frac{95}{37} = 2 + \cfrac{1}{1 + \cfrac{1}{1 + \cfrac{1}{3 + \cfrac{1}{5}}}} \ .$$

This is called the *continued fraction expansion* of $\frac{95}{37}$. To save space we normally use one of two alternative notations:

$$a_0 + \cfrac{1}{a_1 + \cfrac{1}{a_2 + \cdots + \cfrac{1}{a_n}}} = [a_0, a_1, a_2, \ldots, a_n] = a_0 + \frac{1}{a_1 +} \ \frac{1}{a_2 +} \ \cdots \ \frac{1}{a_n} \ .$$

DOI: 10.1201/9781003111207-4

It is easy to see that if $a_0, a_1, a_2, \ldots, a_n$ are positive integers, then the expressions just considered represent rational numbers. If we try to find a continued fraction for an irrational number such as $\sqrt{2}$, we might begin as follows. Since $1 < \sqrt{2} < 2$ we have

$$\sqrt{2} = 1 + (\sqrt{2} - 1) ,$$

and in order to express the "remainder" $\sqrt{2} - 1$ as a fraction with numerator 1 we write

$$\sqrt{2} = 1 + \frac{(\sqrt{2} - 1)(\sqrt{2} + 1)}{\sqrt{2} + 1} = 1 + \frac{1}{\sqrt{2} + 1} .$$

Since $2 < \sqrt{2} + 1 < 3$ we continue

$$\sqrt{2} = 1 + \frac{1}{2 + (\sqrt{2} - 1)} = 1 + \frac{1}{2 +} \frac{1}{\sqrt{2} + 1} ,$$

and we observe that with the reappearance of the denominator $\sqrt{2} + 1$ the process will repeat. Suppressing any qualms about infinite algebraic expressions we might write

$$\sqrt{2} = 1 + \frac{1}{2 +} \frac{1}{2 +} \frac{1}{2 +} \frac{1}{2 +} \cdots ,$$

though we should realise that this raises a convergence question which needs to be resolved.

It's time to introduce some terminology.

4.1 DEFINITION AND BASIC PROPERTIES

Definition 4.1. *A finite or infinite expression of the form*

$$a_0 + \cfrac{b_1}{a_1 + \cfrac{b_2}{a_2 + \cfrac{b_3}{a_3 + \cdots}}}$$

is called a **continued fraction**. *A* **simple** *continued fraction is one in which every b_k is 1, every a_k is an integer, and every a_k except possibly a_0 is positive. For a (finite or infinite) simple continued fraction we shall also use the notations*

$$a_0 + \frac{1}{a_1 +} \frac{1}{a_2 +} \frac{1}{a_3 +} \cdots \qquad \text{and} \qquad [a_0, a_1, a_2, a_3, \ldots] .$$

A finite simple continued fraction is said to **represent** *the number obtained by performing the arithmetic in the obvious way; an infinite simple continued fraction $[a_0, a_1, a_2, a_3, \ldots]$ represents the real number α if*

$$\alpha = \lim_{n \to \infty} [a_0, a_1, a_2, a_3, \ldots, a_n] .$$

*Let $k \in \mathbb{N}$. The integer a_k is called the kth **partial quotient** of the continued fraction $[a_0, a_1, a_2, \ldots]$, or of the number α it represents; the continued fraction $\alpha_k = [a_k, a_{k+1}, a_{k+2}, \ldots]$ is the kth **complete quotient** of α; and the continued fraction $[a_0, a_1, \ldots, a_k]$ is the kth **convergent** to α.*

Note that a convergent, being defined as a finite continued fraction, is always a rational number. Henceforth we shall blur the distinction between a continued fraction and the number represented by the continued fraction. We shall use such language as, for example, "the continued fraction $\alpha = [a_0, a_1, a_2, \ldots]$ " instead of saying more precisely, "the continued fraction $[a_0, a_1, a_2, \ldots]$ which represents the number α".

Continued fractions, their convergents, partial quotients and complete quotients have many fascinating properties.

Lemma 4.1. *Basic properties of continued fractions. Let α be the finite simple continued fraction $[a_0, a_1, \ldots, a_n]$ or the infinite continued fraction $[a_0, a_1, a_2, \ldots]$. Let the kth complete quotient of α be α_k. For integers $k \geq -2$ define p_k and q_k inductively by*

$$p_{-2} = 0, \quad p_{-1} = 1, \quad p_k = a_k p_{k-1} + p_{k-2} \text{ for } k \geq 0;$$
$$q_{-2} = 1, \quad q_{-1} = 0, \quad q_k = a_k q_{k-1} + q_{k-2} \text{ for } k \geq 0.$$

Here and in the following we shall assume in the finite case that $k \leq n$. In any case a list $a_0, a_1, \ldots, a_{k-1}$ with $k = 0$ is taken to be the empty list. We have

- *$\alpha = [a_0, a_1, \ldots, a_{k-1}, \alpha_k]$ for $k \geq 0$;*
- *if $x > 0$ is real and $k \geq 0$, then*

$$[a_0, a_1, \ldots, a_{k-1}, x] = \frac{x p_{k-1} + p_{k-2}}{x q_{k-1} + q_{k-2}};$$

- *if $k \geq 0$ then*

$$\alpha = \frac{\alpha_k p_{k-1} + p_{k-2}}{\alpha_k q_{k-1} + q_{k-2}};$$

- *$p_k / q_k = [a_0, a_1, \ldots, a_k]$ is the kth convergent to α, for $k \geq 0$;*
- *if $k \geq -1$, then*

$$p_{k-1} q_k - p_k q_{k-1} = (-1)^k;$$

- *p_k and q_k are relatively prime for any $k \geq 0$.*

Proof. The first result is clear from the definition. The second is an easy induction proof making use of the identity

$$a_0 + \cfrac{1}{a_1 +} \cfrac{1}{a_2 +} \cdots \cfrac{1}{a_k +} \cfrac{1}{x} = a_0 + \cfrac{1}{a_1 +} \cfrac{1}{a_2 +} \cdots \cfrac{1}{a_{k-1} +} \cfrac{1}{a_k + \frac{1}{x}},$$

and the third follows immediately. Taking $x = a_k$ in the second result yields the fourth, and the fifth is again an induction using

$$p_k q_{k+1} - p_{k+1} q_k = p_k(a_{k+1}q_k + q_{k-1}) - (a_{k+1}p_k + p_{k-1})q_k$$
$$= -(p_{k-1}q_k - p_k q_{k-1}) \,.$$

It follows from the fifth property that any common divisor of p_k and q_k is also a divisor of $(-1)^k$, and this proves the last result.

These properties provide a convenient method for evaluating a finite simple continued fraction α and all its convergents: list the partial quotients a_k and calculate p_k and q_k recursively. Then all the values of p_k/q_k are convergents to α, and the last is α itself. For example, to find the convergents of

$$\alpha = 1 + \cfrac{1}{3+} \, \cfrac{1}{1+} \, \cfrac{1}{5+} \, \cfrac{1}{1+} \, \cfrac{1}{1+} \, \cfrac{1}{4}$$

we construct a table

		1	3	1	5	1	1	4
0	1	1	4	5	29	34	63	286
1	0	1	3	4	23	27	50	227

and then read off the convergents

$$\frac{1}{1}, \; \frac{4}{3}, \; \frac{5}{4}, \; \frac{29}{23}, \; \frac{34}{27}, \; \frac{63}{50}, \; \frac{286}{227},$$

the last of which is the value of α.

We have seen that any finite simple continued fraction represents a rational number. The converse is also true.

Theorem 4.2. *Any rational number is represented by a finite simple continued fraction.*

Proof. Let $\alpha = p/q$ with, as usual, p, q relatively prime and $q > 0$. We shall prove the result by induction on q. First, if $q = 1$ then the continued fraction is just $\alpha = \lfloor p \rfloor$. Let $q > 1$ and assume that any rational number with denominator less than q can be expressed as a finite continued fraction. Divide p by q to give quotient a and remainder r: that is,

$$p = aq + r \quad \text{with} \quad 0 < r < q \,.$$

The remainder r is not zero since p and q are coprime; so by the inductive hypothesis we can write q/r as a finite simple continued fraction $[b_0, b_1, \ldots, b_m]$, and we have

$$\alpha = \frac{p}{q} = a + \frac{r}{q} = a + \frac{1}{q/r} = [a, b_0, b_1, \ldots, b_m] \,.$$

This proves the inductive step, and the result follows.

Although any rational number can be written as a continued fraction, the representation is not unique. For example, we have

$$\frac{95}{37} = 2 + \frac{1}{1+}\ \frac{1}{1+}\ \frac{1}{3+}\ \frac{1}{5} = 2 + \frac{1}{1+}\ \frac{1}{1+}\ \frac{1}{3+}\ \frac{1}{4+}\ \frac{1}{1}\ .$$

We may, if convenient, rule out the second possibility by insisting that the last partial quotient in a continued fraction be greater than 1; and there are no further continued fraction representations of a rational number, as is shown in the following result.

Proposition 4.3. "Almost–uniqueness" of continued fractions. *Suppose that* $[a_0, a_1, \ldots, a_m]$ *and* $[b_0, b_1, \ldots, b_n]$ *are finite simple continued fractions representing the same (rational) number, where by symmetry we may assume that* $0 \leq m \leq n$. *Then either*

- $m = n$ *and* $a_0 = b_0$, $a_1 = b_1, \ldots,$ $a_m = b_m$; *or*
- $m = n - 1$ *and* $a_0 = b_0$, $a_1 = b_1, \ldots,$ $a_{m-1} = b_{m-1}$, *with* $a_m = b_m + 1$ *and* $b_n = 1$.

Proof. First, consider the case $m = 0$; then $a_0 = [b_0, b_1, \ldots, b_n]$. If $n = 0$, then $a_0 = b_0$. If $n = 1$, then $a_0 = b_0 + 1/b_1$ and so $b_0 < a_0 \leq b_0 + 1$; since a_0 and b_0 are integers, we have $a_0 = b_0 + 1$ and the second alternative above holds. If $n > 1$, then

$$a_0 = b_0 + \cfrac{1}{b_1 + \cfrac{1}{b_2}}\ ;$$

but $b_1 + 1/b_2$ is strictly greater than 1, so $b_0 < a_0 < b_0 + 1$, which is impossible. In the case $m \geq 1$ we write

$$\alpha = [a_0, a_1, \ldots, a_m] = [b_0, b_1, \ldots, b_n]\ ;$$

reasoning as in the case $n = 1$ above, we have

$$a_0 < \alpha \leq a_0 + 1 \quad \text{and} \quad b_0 < \alpha \leq b_0 + 1\ .$$

Now if α is an integer then $a_0 + 1 = \alpha = b_0 + 1$, while if not then $a_0 = \lfloor \alpha \rfloor = b_0$; but in either case, $a_0 = b_0$. Therefore

$$a_0 + \frac{1}{[a_1, \ldots, a_m]} = a_0 + \frac{1}{[b_1, \ldots, b_n]}\ ,$$

so $[a_1, \ldots, a_m] = [b_1, \ldots, b_n]$ and the result follows by induction.

4.2 CONTINUED FRACTIONS OF IRRATIONAL NUMBERS

Consider again our (presumed) continued fraction for $\sqrt{2}$. Using the tabular method, or otherwise, we find that the first few convergents to $\sqrt{2}$ are

$$\frac{1}{1}, \frac{3}{2}, \frac{7}{5}, \frac{17}{12}, \frac{41}{29}, \frac{99}{70}, \frac{239}{169}, \frac{577}{408}, \ldots$$

Evaluating these convergents (and, if necessary, a few more) as decimals, it is not hard to convince ourselves that the convergents p_{2k}/q_{2k} with even indices form an increasing sequence converging to the limit $\sqrt{2}$, while the convergents p_{2k+1}/q_{2k+1} with odd indices form a decreasing sequence which converges to the same limit. In fact, this observation is the key to proving that an infinite simple continued fraction always converges; having done so, we shall find it easy to confirm that the continued fraction for $\sqrt{2}$ is as we have conjectured.

Lemma 4.4. Oscillation of convergents. *Let p_k/q_k be the kth convergent to the infinite simple continued fraction $\alpha = [a_0, a_1, a_2, \ldots]$. Then*

$$\frac{p_0}{q_0} < \frac{p_2}{q_2} < \frac{p_4}{q_4} < \cdots < \frac{p_5}{q_5} < \frac{p_3}{q_3} < \frac{p_1}{q_1} . \tag{4.1}$$

Moreover, q_k increases without limit as $k \to \infty$.

Proof. It is obvious that $p_0/q_0 < p_1/q_1$. For any $k \geq 2$, we have

$$\frac{p_k}{q_k} = \frac{a_k p_{k-1} + p_{k-2}}{a_k q_{k-1} + q_{k-2}} ;$$

since all terms involved are positive, a result from elementary algebra (see appendix 1) shows that p_k/q_k lies between p_{k-1}/q_{k-1} and p_{k-2}/q_{k-2}. Applying this result repeatedly proves (4.1). To prove the second part of the lemma, note first that $q_0 = 1$ and $q_1 = a_1 \geq 1$; then for $k \geq 2$, we have

$$q_k = a_k q_{k-1} + q_{k-2} \geq q_{k-1} + q_{k-2} \geq q_{k-1} + 1 ,$$

and so $q_k \to \infty$ as $k \to \infty$.

Comment. If $\alpha = [a_0, a_1, \ldots, a_n]$ is a finite continued fraction, then equation (4.1) still holds provided that we omit all p_k/q_k with $k > n$.

Theorem 4.5. *Any infinite simple continued fraction converges to a limit; moreover, this limit is irrational.*

Proof. Let p_k/q_k be the kth convergent to $[a_0, a_1, a_2, \ldots]$. From the lemma,

$$\left\{ \frac{p_0}{q_0}, \frac{p_2}{q_2}, \frac{p_4}{q_4}, \ldots \right\}$$

is a monotonic increasing sequence which is bounded above (for example, by p_1/q_1), and therefore tends to a limit α_L with $\alpha_L \geq p_{2k}/q_{2k}$ for all k. Similarly

$$\left\{ \frac{p_1}{q_1}, \frac{p_3}{q_3}, \frac{p_5}{q_5}, \ldots \right\}$$

decreases to a limit α_U with $\alpha_U \leq p_{2k+1}/q_{2k+1}$ for all k. Moreover $\alpha_U \geq \alpha_L$, for otherwise we should be able to find convergents p_{2k}/q_{2k} arbitrarily close to α_L and $p_{2\ell+1}/q_{2\ell+1}$ arbitrarily close to α_U; but then $p_{2\ell+1}/q_{2\ell+1}$ would be smaller than p_{2k}/q_{2k}, contradicting (4.1). Therefore, for any k, we have

$$0 \leq \alpha_U - \alpha_L \leq \frac{p_{2k+1}}{q_{2k+1}} - \frac{p_{2k}}{q_{2k}} = \frac{1}{q_{2k+1}q_{2k}} .$$

However, $\alpha_U - \alpha_L$ is independent of k, and from the lemma we know that $1/q_{2k+1}q_{2k} \to 0$ as $k \to \infty$; so $\alpha_U = \alpha_L = \alpha$, say. Thus $[a_0, a_1, \ldots, a_k]$ tends to the limit α as $k \to \infty$, and this means by definition that the continued fraction represents the real number α.

To prove the irrationality of α we use the test provided by Lemma 3.18. If α is rational there is a positive constant c such that

$$\left| \alpha - \frac{p_{2k}}{q_{2k}} \right| \geq \frac{c}{q_{2k}}$$

for all convergents p_{2k}/q_{2k}, with at most one exception. But for every k, we have

$$0 \leq \alpha - \frac{p_{2k}}{q_{2k}} \leq \frac{p_{2k+1}}{q_{2k+1}} - \frac{p_{2k}}{q_{2k}} = \frac{1}{q_{2k+1}q_{2k}} .$$

Combining these estimates yields $q_{2k+1} < 1/c$ for all (except possibly one) k, which is not true as the denominators q_{2k+1} are unbounded.

Comment. The latter part of this proof is *not* redundant. We showed in Theorem 4.2 above that a rational number has a finite continued fraction representation (in fact, two of them); but we did not prove that a rational has *only* finite representations.

There is essentially only one fact we have yet to prove concerning the existence and uniqueness (or not) of the continued fraction for a given real number.

Theorem 4.6. *If α is a real irrational number, then it has a unique expansion as a simple continued fraction (and we already know that this continued fraction must be infinite).*

Proof. Following the procedure we used for $\sqrt{2}$ on page 66, we set $\alpha_0 = \alpha$ and then for $k \geq 0$ define recursively

$$a_k = \lfloor \alpha_k \rfloor \quad \text{and} \quad \alpha_{k+1} = \frac{1}{\alpha_k - a_k} . \tag{4.2}$$

Then a_0 is an integer, $a_1, a_2, \ldots$ are positive integers and $\alpha_1, \alpha_2, \ldots$ are real numbers exceeding 1. Since α is irrational we can readily show that every α_k is irrational; so $\alpha_k - a_k$ cannot vanish and the recursion never terminates.

Therefore, we have an infinite simple continued fraction, which by the previous theorem converges to a limit, say

$$\beta = a_0 + \cfrac{1}{a_1 +} \cfrac{1}{a_2 +} \cfrac{1}{a_3 +} \cdots .$$

We have to show that $\alpha = \beta$. Now from (4.2) we prove by induction that

$$\alpha = [a_0, a_1, \ldots, a_{k-1}, \alpha_k] \tag{4.3}$$

for all k, and hence by a property from Lemma 4.1,

$$\alpha = \frac{\alpha_k p_{k-1} + p_{k-2}}{\alpha_k q_{k-1} + q_{k-2}} .$$

Here $p_k/q_k = [a_0, a_1, \ldots, a_k]$ is the kth convergent to β. Note that we cannot say immediately that p_k/q_k is a convergent to α because the expression (4.3) is not a *simple* continued fraction. Using appendix 1 again, α lies between

$$\frac{\alpha_k p_{k-1}}{\alpha_k q_{k-1}} \quad \text{and} \quad \frac{p_{k-2}}{q_{k-2}} .$$

But as $k \to \infty$ each of these fractions tends to the limit β, and so $\alpha = \beta$ as required.

To prove uniqueness, suppose that we have two infinite simple continued fractions representing the same number,

$$[a_0, a_1, a_2, \ldots] = [b_0, b_1, b_2, \ldots] .$$

For any k, we can write these in the form

$$[a_0, a_1, \ldots, a_k, \alpha_{k+1}] = [b_0, b_1, \ldots, b_k, \beta_{k+1}]$$

with $\alpha_{k+1}, \beta_{k+1} > 1$, all a_j, b_j integers and all except possibly a_0, b_0 positive. Then from the following lemma (which will be useful again later), we have immediately $a_k = b_k$. As this is true for all k, the two continued fractions are identical.

Lemma 4.7. A uniqueness property. *Suppose that $0 \le m \le n$ and*

$$[a_0, a_1, \ldots, a_{m-1}, \alpha_m] = [b_0, b_1, \ldots, b_{n-1}, \beta_n] ,$$

where a_0, b_0 are integers; $a_1, \ldots, a_{m-1}, b_1, \ldots, b_{n-1}$ are positive integers; and α_m, β_n are real numbers strictly greater than 1. Then

$$a_0 = b_0 , \quad a_1 = b_1 , \ldots, \quad a_{m-1} = b_{m-1} \quad \text{and} \quad \alpha_m = [b_m, \ldots, b_{n-1}, \beta_n] .$$

Comment. This result can be expressed as follows: if two continued fractions, each ending with complete quotients greater than 1, represent the same number, then "as far as they are both simple" they are identical.

Proof. Let $m \geq 1$ and suppose that

$$\alpha = [a_0, a_1, \ldots, a_{m-1}, \alpha_m] = [b_0, b_1, \ldots, b_{n-1}, \beta_n]$$

with the conditions of the theorem satisfied. Observe that

$$[a_1, \ldots, a_{m-1}, \alpha_m] > 1 ;$$

for if $m = 1$ (so that there are no terms a_k) then the left–hand side is just α_m, which is greater than 1 by assumption, while if $m > 1$ then the left–hand side is the sum of a positive integer a_1 and a strictly positive quantity. Hence

$$0 < \cfrac{1}{a_1 +} \cdots \cfrac{1}{a_{m-1} +} \cfrac{1}{\alpha_m} < 1$$

and we have $a_0 < \alpha < a_0 + 1$. Similar reasoning holds for the other continued fraction, and so

$$a_0 = \lfloor \alpha \rfloor = b_0 .$$

Consequently

$$[a_1, \ldots, a_{m-1}, \alpha_m] = [b_1, \ldots, b_{n-1}, \beta_n] ,$$

and the result will follow from the corresponding result for $m - 1$ terms. Since the case $m = 0$ is vacuously true (almost), the result is proved by mathematical induction.

Examples. Although we now know that $\sqrt{2}$ has a continued fraction, and that $[1, 2, 2, 2, \ldots]$ converges, we still have not shown that they are equal! To do this, write

$$\alpha = 1 + \cfrac{1}{2 +} \cfrac{1}{2 +} \cfrac{1}{2 +} \cdots \quad \text{and} \quad \beta = 2 + \cfrac{1}{2 +} \cfrac{1}{2 +} \cdots .$$

Then $\beta = 2 + 1/\beta$, which simplifies to $\beta^2 - 2\beta - 1 = 0$. Since β is positive we have $\beta = 1 + \sqrt{2}$; but $\alpha = 1 + 1/\beta$ and at last we have shown properly that

$$\sqrt{2} = 1 + \cfrac{1}{2 +} \cfrac{1}{2 +} \cfrac{1}{2 +} \cdots .$$

As a second example we evaluate $\alpha = [5, 4, 3, 2, 1, 3, 2, 1, \ldots]$, a *periodic* continued fraction. Set $\beta = [3, 2, 1, 3, 2, 1, \ldots]$; then

$$\alpha = 5 + \cfrac{1}{4 +} \cfrac{1}{\beta} \quad \text{and} \quad \beta = 3 + \cfrac{1}{2 +} \cfrac{1}{1 +} \cfrac{1}{\beta} .$$

Evaluating the latter fraction by the tabular method

		3	2	1	β
0	1	3	7	10	$10\beta + 7$
1	0	1	2	3	$3\beta + 2$

gives

$$\beta = \frac{10\beta + 7}{3\beta + 2} \ ,$$

which leads to the quadratic equation $3\beta^2 - 8\beta - 7 = 0$ with positive root $\beta = \frac{1}{3}(4 + \sqrt{37})$; using similar means to evaluate α in terms of β, we find that

$$5 + \frac{1}{4+} \ \frac{1}{3+} \ \frac{1}{2+} \ \frac{1}{1+} \ \frac{1}{3+} \ \frac{1}{2+} \ \frac{1}{1+} \ \cdots \ = \frac{21\beta + 5}{4\beta + 1} = \frac{409 - \sqrt{37}}{77} \ .$$

In fact the same procedure will show that any (eventually) periodic continued fraction is a quadratic irrational. Consider

$$\alpha = [a_0, a_1, \ldots, a_{m-1}, b_0, b_1, \ldots, b_{n-1}, b_0, b_1, \ldots, b_{n-1}, b_0, \ldots] \ ,$$

and let

$$\beta = [b_0, b_1, \ldots, b_{n-1}, b_0, b_1, \ldots, b_{n-1}, b_0, \ldots] \ .$$

If we write p_k/q_k for the convergents to α and r_k/s_k for the convergents to β, we have

$$\alpha = [a_0, a_1, \ldots, a_{m-1}, \beta] = \frac{\beta \, p_{m-1} + p_{m-2}}{\beta \, q_{m-1} + q_{m-2}}$$

and

$$\beta = [b_0, b_1, \ldots, b_{n-1}, \beta] = \frac{\beta \, r_{n-1} + r_{n-2}}{\beta \, s_{n-1} + s_{n-2}} \ .$$

The second equation reduces to a quadratic with integral coefficients, so β is a quadratic irrational (the quadratic cannot have rational roots since β has an infinite continued fraction). Substituting the value of β into the previous equation and, if desired, rationalising the denominator shows that α is also a quadratic irrational. We have proved the following result.

Theorem 4.8. *An eventually periodic simple continued fraction represents a quadratic irrational.*

In fact, the converse of this result is true, so that a quadratic irrational is characterised by having an (infinite) eventually periodic continued fraction. We shall prove this immediately in order to complete our study of quadratic irrationals, even though the proof requires a result concerning approximation properties of convergents that we shall not prove until later. We leave it to the reader to confirm that our proof of Lemma 4.10 does not depend on any result which has not been proved up to this point.

Theorem 4.9. *The continued fraction of a quadratic irrational number is eventually periodic.*

Proof. Let α be a root of the (irreducible) quadratic $Az^2 + Bz + C$, where A, B and C are integers with A and C not zero. Since the convergents and the complete quotients of α satisfy the relation

$$\alpha = \frac{\alpha_k p_{k-1} + p_{k-2}}{\alpha_k q_{k-1} + q_{k-2}}$$

we have

$$A\left(\frac{\alpha_k p_{k-1} + p_{k-2}}{\alpha_k q_{k-1} + q_{k-2}}\right)^2 + B\left(\frac{\alpha_k p_{k-1} + p_{k-2}}{\alpha_k q_{k-1} + q_{k-2}}\right) + C = 0 \; ;$$

once we emerge from algebraic manipulations we find that α_k is a root of a quadratic $A_k z^2 + B_k z + C_k$, where A_k, B_k and C_k are integers given by

$$A_k = A p_{k-1}^2 + B p_{k-1} q_{k-1} + C q_{k-1}^2$$
$$B_k = 2A p_{k-1} p_{k-2} + B p_{k-1} q_{k-2} + B p_{k-2} q_{k-1} + 2C q_{k-1} q_{k-2}$$
$$C_k = A p_{k-2}^2 + B p_{k-2} q_{k-2} + C q_{k-2}^2 \; .$$

Now write $d_k = q_k \alpha - p_k$; by the inequality in (4.5), we have

$$|d_k| = q_k \left|\alpha - \frac{p_k}{q_k}\right| < \frac{1}{q_{k+1}} \leq \frac{1}{q_k} \leq 1 \; . \tag{4.4}$$

Therefore

$$\begin{aligned} A_k &= A(q_{k-1}\alpha - d_{k-1})^2 + B(q_{k-1}\alpha - d_{k-1})q_{k-1} + C q_{k-1}^2 \\ &= (A\alpha^2 + B\alpha + C)q_{k-1}^2 - (2A\alpha + B)d_{k-1}q_{k-1} + A d_{k-1}^2 \\ &= A d_{k-1}^2 - (2A\alpha + B)d_{k-1}q_{k-1} \; , \end{aligned}$$

and using the inequality (4.4) we obtain

$$|A_k| \leq |A| + |2A\alpha + B| \; .$$

Performing very similar calculations for B_k and observing that

$$q_{k-1}d_{k-2} + q_{k-2}d_{k-1} \leq \frac{q_{k-1}}{q_{k-1}} + \frac{q_{k-2}}{q_k} < 2$$

we find that

$$|B_k| \leq 2(|A| + |2A\alpha + B|) \; ;$$

finally,

$$|C_k| = |A_{k-1}| \leq |A| + |2A\alpha + B| \; .$$

Hence the coefficients A_k, B_k and C_k are bounded independently of k; so they take only finitely many values, and the quadratics $A_k z^2 + B_k z + C_k$ have altogether only finitely many roots. Therefore, we must have $\alpha_k = \alpha_{k+t}$ for some k; this implies that

$$a_k = \lfloor \alpha_k \rfloor = \lfloor \alpha_{k+t} \rfloor = a_{k+t} \; ,$$

and that the same relations must hold for all subsequent k. Hence the continued fraction of α is eventually periodic.

Comments.

- As often happens with this kind of argument, the number of possibilities for α_k given by our estimate is far in excess of the actual number. For example, if $\alpha = \sqrt{2}$ then we have $A = 1$, $B = 0$ and $C = -2$, which leads to

$$0 < |A_k| \leq 3 \,, \quad |B_k| \leq 7 \quad \text{and} \quad 0 < |C_k| \leq 3 \,.$$

Even if we assume (as we may) that A_k is positive, this gives 270 possible quadratics and 540 possible α_k. But in fact α_k takes only two different values!

- An alternative argument for part of the above proof: by straightforward algebra, $B_k^2 - 4A_kC_k = B^2 - 4AC$, which is a fixed constant. So once we have proved that A_k and C_k take only finitely many values, the same follows immediately for B_k.

- Why does this not work for irrationalities of higher degree? Well, if α is (say) a cubic irrational, we should expect to get a formula something like

$$\begin{aligned} A_k = \,&(A\alpha^3 + B\alpha^2 + C\alpha + D)q_{k-1}^3 \\ &- (3A\alpha^2 + 2B\alpha + C)q_{k-1}^2 d_{k-1} \\ &+ (3A\alpha + B)q_{k-1}d_{k-1}^2 - Ad_{k-1}^3 \,. \end{aligned}$$

Now as in the quadratic case, the first term here will vanish and the last two will be bounded. However, the second term will be, roughly, a constant times q_{k-1}, which is unbounded, and this will wreck the entire argument.

- In view of the theorem just proved and earlier results, we can sum up the connection between the continued fraction of a real number and its algebraic status in this way: algebraic numbers of degree 1 have finite continued fractions; those of degree 2 have ultimately periodic infinite continued fractions; and those of higher degree, as well as transcendental numbers, have non–periodic infinite continued fractions.

4.3 APPROXIMATION PROPERTIES OF CONVERGENTS

Having proved all that we need about representation of numbers by continued fractions, we proceed to investigate what continued fractions can tell us about the approximation of irrationals by rationals. This will link the present topic with that of the previous chapter. First, some equalities and inequalities concerning the difference between a number and its convergents.

Lemma 4.10. *Let* $\alpha = [a_0, a_1, a_2, \ldots]$ *be an infinite simple continued fraction with convergents* p_k/q_k *and complete quotients* α_k. *Then for* $k \geq 0$ *we have*

$$\left| \alpha - \frac{p_k}{q_k} \right| = \frac{1}{(\alpha_{k+1}q_k + q_{k-1})q_k} < \frac{1}{q_{k+1}q_k} \leq \frac{1}{a_{k+1}q_k^2} \,. \tag{4.5}$$

If $\alpha = [a_0, a_1, \ldots, a_n]$ is a finite continued fraction, then the same relations hold for $0 \le k < n - 1$, while if $k = n - 1$ the first inequality must be replaced by equality.

Proof. We have

$$\alpha - \frac{p_k}{q_k} = \frac{\alpha_{k+1}p_k + p_{k-1}}{\alpha_{k+1}q_k + q_{k-1}} - \frac{p_k}{q_k} = \frac{p_{k-1}q_k - p_k q_{k-1}}{(\alpha_{k+1}q_k + q_{k-1})q_k},$$

and the equality follows since $p_{k-1}q_k - p_k q_{k-1} = \pm 1$. To prove the inequalities we need only observe that, except in the finite case when $k + 1 = n$, we have

$$\alpha_{k+1}q_k + q_{k-1} > a_{k+1}q_k + q_{k-1} = q_{k+1} \ge a_{k+1}q_k \; .$$

Comments.

- Using this we can give an alternative proof of Theorem 3.21, that any irrational is approximable to order 2. For an irrational number α has infinitely many convergents p/q, and, from (4.5), every one of these satisfies

$$0 < \left| \alpha - \frac{p}{q} \right| < \frac{1}{q^2} \; .$$

- Observe also from (4.5) that p_k/q_k will approximate α to within c/q_k^2, and that the constant c will be exceptionally small if the *next* partial quotient a_{k+1} is large. Now consider the continued fraction for π. By taking a sufficiently accurate decimal approximation to π, or possibly by other methods, we obtain

$$\pi = 3 + \frac{1}{7+} \; \frac{1}{15+} \; \frac{1}{1+} \; \frac{1}{292+} \; \frac{1}{1+} \; \frac{1}{1+} \; \frac{1}{1+} \; \frac{1}{2+} \cdots .$$

So we can find two very good fractional approximations to π by truncating the continued fraction just before the partial quotients 15 and 292. The first of these gives $\pi \approx [3, 7] = \frac{22}{7}$. This approximation is, of course, well known; perhaps less well known is just how good an approximation it is. The discrepancy is

$$\left| \pi - \frac{22}{7} \right| < \frac{1}{15 \times 7^2} = \frac{1}{735} \; ;$$

to obtain similar accuracy from the decimal expansion we would need to take two decimal places, and this would mean, in effect, to approximate π by $\frac{314}{100}$, a much more complicated fraction than $\frac{22}{7}$. The other approximation cited above gives

$$\pi \approx 3 + \frac{1}{7+} \; \frac{1}{15+} \; \frac{1}{1} = \frac{355}{113} \quad \text{and} \quad \left| \pi - \frac{355}{113} \right| < \frac{1}{292 \times 113^2} \; .$$

This estimate was known to the Chinese mathematician Zu Chongzhi (A.D. 429–501); it is accurate to six decimal places, and is a closer approximation than the fraction $\frac{3141593}{1000000}$.

We shall use the inequalities of the preceding lemma to look more closely at the way in which numbers are approximated by their convergents, and thereby to see how the continued fraction of a real number α is related to its order of approximation by rationals. Since we already know (pages 46–47) everything there is to know about approximability properties of rationals, we shall generally assume that α is irrational, and shall restrict ourselves to occasional comments on the rational case. Many of the properties we shall prove do in fact also apply to this case, with minor modifications sometimes being necessary. First, we show that the discrepancy between α and a convergent p_k/q_k always decreases as k increases.

Lemma 4.11. *Let* $\alpha = [a_0, a_1, a_2, \ldots]$ *be an infinite simple continued fraction with convergents* p_k/q_k *and complete quotients* α_k. *Then for* $k \geq 1$, *we have*

$$|q_k\alpha - p_k| < |q_{k-1}\alpha - p_{k-1}| \quad and \quad \left|\alpha - \frac{p_k}{q_k}\right| < \left|\alpha - \frac{p_{k-1}}{q_{k-1}}\right| .$$

Proof. We use the equality in (4.5). The first inequality to be proved is equivalent to

$$\alpha_{k+1}q_k + q_{k-1} > \alpha_k q_{k-1} + q_{k-2} ;$$

substituting $q_{k-2} = q_k - a_k q_{k-1}$ and rearranging shows that we need to prove

$$(\alpha_{k+1} - 1)q_k > (\alpha_k - a_k - 1)q_{k-1} .$$

But for an infinite continued fraction it is always true that $\alpha_{k+1} > 1$, and hence that $\alpha_k - a_k = 1/\alpha_{k+1} < 1$; therefore

$$(\alpha_{k+1} - 1)q_k > 0 \geq (\alpha_k - a_k - 1)q_{k-1} ,$$

and the first part of the proof is complete. The second inequality follows immediately since $q_k \geq q_{k-1}$.

Comment. If $\alpha = [a_0, a_1, \ldots, a_n] \in \mathbb{Q}$ and if $a_n > 1$, then the inequalities just proved hold for $1 \leq k \leq n$. If $a_n = 1$ and $k = n - 1$ the first inequality is false, and must be replaced by an equality. However, we may ignore this case, since, as pointed out on page 69, a rational number can (nearly) always be written as a continued fraction whose last partial quotient is strictly greater than 1.

The following result shows that the "best" rational approximations to an irrational number are its convergents, as foreshadowed at the beginning of this chapter.

Theorem 4.12. **Convergents are best approximations.** *Let* α *be a real irrational with convergents* p_k/q_k.

- *If* p/q *is a rational number (with, as usual, positive denominator) such that*

$$\left|\alpha - \frac{p}{q}\right| < \left|\alpha - \frac{p_k}{q_k}\right| \tag{4.6}$$

  with $k \geq 1$, *then* $q > q_k$.

- *If p and q are integers, $q > 0$, such that*

$$|q\alpha - p| < |q_k\alpha - p_k| \qquad (4.7)$$

  with $k \geq 0$, then $q \geq q_{k+1}$.

Proof. We shall prove the second part of the theorem first, and then show that the first part easily follows. So, suppose that α is irrational, that $k \geq 0$ and that $|q\alpha - p| < |q_k\alpha - p_k|$; consider the system of linear equations

$$\begin{cases} p_k x + p_{k+1} y = p \\ q_k x + q_{k+1} y = q \, . \end{cases}$$

The determinant of the coefficients on the left–hand side is $p_k q_{k+1} - p_{k+1} q_k$, which is ± 1, and so the system has a solution in integers x, y (see the second appendix to this chapter). Recall that q, q_k and q_{k+1} are all positive; from the second equation above it is therefore clear that x and y cannot both be negative, while if both are positive then $q > q_{k+1} y \geq q_{k+1}$ and the result is proved. So we may assume that one of x, y is positive (or zero), and the other is negative (or zero). It follows from the proof of Theorem 4.5 that one of

$$\alpha - \frac{p_k}{q_k} \quad \text{and} \quad \alpha - \frac{p_{k+1}}{q_{k+1}}$$

is positive, one negative; and so the same is true for the expressions $q_k\alpha - p_k$ and $q_{k+1}\alpha - p_{k+1}$. Therefore

$$x(q_k\alpha - p_k) \quad \text{and} \quad y(q_{k+1}\alpha - p_{k+1})$$

have the same sign, and hence

$$|x|\,|q_k\alpha - p_k| + |y|\,|q_{k+1}\alpha - p_{k+1}| = |x(q_k\alpha - p_k) + y(q_{k+1}\alpha - p_{k+1})| \, .$$

But the right–hand side is just $|q\alpha - p|$, and so

$$|x|\,|q_k\alpha - p_k| \leq |q\alpha - p| < |q_k\alpha - p_k| \, .$$

This can be true only if $x = 0$, so $q = q_{k+1} y \geq q_{k+1}$ and the second result stated above is proved.

To prove the first result, suppose that the inequality (4.6) holds for some $k \geq 1$ but that $q \leq q_k$. Multiplying these inequalities (permissible as all the quantities involved are positive) we obtain $|q\alpha - p| < |q_k\alpha - p_k|$. From the result already proved, we have $q \geq q_{k+1}$, and so $q_{k+1} \leq q_k$, which is not possible for $k \geq 1$.

Comments.

- The first result shows that if p/q is a better approximation to α than p_k/q_k, then the former fraction must have a larger denominator than the latter; in other words, of all fractions with denominators up to q_k, the convergent p_k/q_k is the closest to α. In fact, if we wish to obtain a better approximation than p_k/q_k in the stronger sense (4.7), we have to go at least as far as the *next* convergent.

- The condition $k \geq 1$ in the first result is necessary because $q_0 = q_1$ for some continued fractions α. In fact, if $n + \frac{1}{2} < \alpha < n + 1$ then we have

$$\alpha = n + \cfrac{1}{1 + \cdots} \quad \text{and} \quad \frac{p_0}{q_0} = \frac{n}{1} .$$

However, if $p/q = (n+1)/1$ then the inequality (4.6) is satisfied but the conclusion $q > q_0$ is false.

- An interesting alternative approach to continued fractions is to *define* the convergents to a real number α as the best rational approximations to α, and then derive all of the important properties of convergents. Specifically, we let $p_0 = \lfloor \alpha \rfloor$ and $q_0 = 1$, and then define, for each $k \geq 0$, the convergent p_{k+1}/q_{k+1} to be the fraction with smallest possible denominator such that

$$|q_{k+1}\alpha - p_{k+1}| < |q_k\alpha - p_k| .$$

We can then show that q_{k-1} is a factor of $q_k - q_{k-2}$, and define the partial quotients of α by $a_k = (q_k - q_{k-2})/q_{k-1}$. For more on this approach see Cassels [17].

The next result follows easily from that above.

Theorem 4.13. "Best approximations are convergents". *Let α be irrational. If*

$$\left| \alpha - \frac{p}{q} \right| < \frac{1}{2q^2}$$

then p/q is a convergent to α.

Comment. Conversely, it can be shown that of any two consecutive convergents to α, at least one satisfies the above inequality. We leave this as an exercise.

Proof. Suppose that the given inequality holds. Since $q_k \to \infty$ as $k \to \infty$, there exists $k \geq 0$ such that $q_k \leq q < q_{k+1}$. From the previous theorem we have

$$|q_k\alpha - p_k| \leq |q\alpha - p|$$

because $q < q_{k+1}$, and hence

$$|pq_k - p_kq| \leq q|q_k\alpha - p_k| + q_k|q\alpha - p| \leq (q + q_k)|q\alpha - p| < \frac{q + q_k}{2q} \leq 1 .$$

Since the left–hand side is an integer it must be zero, and $p/q = p_k/q_k$, which is a convergent to α.

4.4 TWO IMPORTANT APPROXIMATION PROBLEMS

The above theorem shows, roughly speaking, that convergents are not merely good approximations but in a sense *the only possible* good approximations to a real number. We digress to look at two important problems of rational approximation which may be solved using continued fractions.

4.4.1 How many days should we count in a calendar year?

This problem was addressed by Euler in [25]. The difficulty is that for convenience of use, the calendar really should contain an integral number of days per year, whereas the actual length of a solar year (that is, the period from one northern spring equinox to the next) can be measured as 365 days, 5 hours, 48 minutes and 46 seconds. If a year were to contain an exact number of days, and always the same number, the calendar would not keep pace with the seasons; students who like to finish their exams around the beginning of December and then head for the beach[1] would at some (not very distant!) date find the weather in December rather unsuitable for this. The main impetus for the creation of the modern calendar came in the mediæval period, when it was realised that accumulated errors in the calendar would eventually lead to Easter, traditionally a spring festival in the northern hemisphere, being celebrated in midwinter.

The solution to the calendar problem is, in outline, simple and well known: we adopt 365 days as the standard length of a year, and decree that certain *leap years* shall be allocated one extra day. The difficulty lies in the details. How shall we determine precisely which years are to be leap years?

Suppose that in a cycle of q years we add an extra day in each of p years. Then the average length of a calendar year will be $365 + p/q$ days, and we would like this to equal the observed length of a solar year. Converting the above data into a rational number of days, and deducting 365, we want

$$\frac{p}{q} = \frac{10463}{43200}.$$

While we could obtain an exact fit to the observations by choosing 10463 years in every 432 centuries as leap years, and then repeating the pattern, it is clear that such a scheme would be too cumbersome for practical use. What we need, therefore, is a good approximation to p/q having a much smaller denominator; and as we have seen, this is a question which can be answered by examining the convergents of p/q. We may compute the continued fraction

$$\frac{10463}{43200} = \frac{1}{4+}\ \frac{1}{7+}\ \frac{1}{1+}\ \frac{1}{3+}\ \frac{1}{5+}\ \frac{1}{64},$$

---

[1]Note for northern hemisphere readers: this was written in Australia, where the academic teaching year traditionally runs from March to November. The summer months are December, January and February.

from which we find the convergents

$$\frac{1}{4}, \quad \frac{7}{29}, \quad \frac{8}{33}, \quad \frac{31}{128}, \quad \frac{163}{673} \quad \text{and} \quad \frac{10463}{43200}.$$

The first of these convergents suggests that we make every fourth year a leap year. This very simple rule constitutes the *Julian calendar*, so called because it was instituted by Julius Caesar, on the advice of the Alexandrian astronomer Sosigenes, in about 45 BC. According to Sosigenes' scheme, four calendar years would be about 45 minutes longer than four solar years, and so, for example, midsummer's day would occur slightly earlier (according to the calendar) every four years. The dates of midsummer and midwinter would be interchanged after about 23000 years; while this prospect would not appear close enough to worry anyone, smaller but still significant discrepancies became noticeable within a few centuries, and led to further reforms of the calendar.

We shall ignore the second convergent $\frac{7}{29}$; it is not much simpler than the following one, $\frac{8}{33}$, and it is not an exceptionally good approximation anyway. We recall that this is because the *third* partial quotient here is not very large. The third convergent can be used to improve on the Julian calendar. It prompts us to declare 8 leap years in 33 years, or 24 in 99 years. Let us adjust this to 24 leap years in 100 years. The approximation will then be less good, but it will be very easy to implement. We need only decree that every fourth year shall be a leap year, with the exception of one year per century, say the century year itself, which shall be an ordinary 365–day year. Using this calendar it would take over 80000 years for the dates of midsummer and midwinter to be interchanged.

The Julian calendar was in use throughout Europe for 1600 years or so, by which time the difference between the calendar year and the solar year was causing perceptible problems. The difficulty was not specifically with the alteration of the seasons but with calculating the date of Easter. This date depends on the first occurrence of the full moon after the (northern) spring equinox. Since the Julian calendar year is longer than the true solar year, the assumed date, March 21, for the equinox will bit by bit arrive later than the true date. Even an error of a day or two – and by the 1500s the accumulated error was about ten days – could cause the first full moon after the equinox to be missed, and a "wrong" full moon to be chosen instead. As a consequence Easter could be celebrated a month or more later than it should have been, and in the late sixteenth century Pope Gregory XIII instituted a commission to advise on how to correct the accumulated errors of 1600 years and minimise future errors. The scheme proposed by this commission, adopted in 1582 in most European Catholic countries and nowadays standard throughout the world, is known as the *Gregorian calendar*.

Consider the fourth convergent $\frac{31}{128}$ to our required ratio. In itself this would suggest a 128–year cycle of ordinary and leap years, which probably would not be convenient to use. But we might notice that

$$\frac{31}{128} = \frac{96\frac{7}{8}}{400} \approx \frac{97}{400} .$$

To employ a scheme based on this approximation we begin by decreeing a leap year every four years, but then omit three of these leap years – that is, restore them to 365 days – every 400 years. The precise method in use today is that every year divisible by 4 is a leap year, except that a year divisible by 100 is not a leap year, except that a year divisible by 400 once again is a leap year. The average length of a year under the Gregorian calendar is 365 days, 5 hours, 49 minutes and 12 seconds, just 26 seconds longer than the solar year.

There is another way to derive this approximation from the convergents to p/q. If p_{k-1}/q_{k-1} and p_k/q_k are adjacent convergents to any number α, consider the fractions

$$\frac{p_{k-1}}{q_{k-1}}, \quad \frac{p_{k-1} + p_k}{q_{k-1} + q_k}, \quad \frac{p_{k-1} + 2p_k}{q_{k-1} + 2q_k}, \quad \frac{p_{k-1} + 3p_k}{q_{k-1} + 3q_k}, \dots, \frac{p_{k-1} + a_{k+1}p_k}{q_{k-1} + a_{k+1}q_k} .$$

The last of these is the convergent p_{k+1}/q_{k+1}; the first is obviously a convergent; the others are known as *secondary convergents*. Roughly speaking, the secondary convergents to α are, after the convergents, the next best rational approximations to α. Here one of the secondary convergents to $\frac{10463}{43200}$ is

$$\frac{31 + 163}{128 + 673} = \frac{194}{801} \approx \frac{194}{800} = \frac{97}{400} ,$$

from which we obtain again the Gregorian approximation.

Many documents from Pope Gregory's scholars still exist, and it is clear that they did *not* in fact use continued fractions to derive their calendar. According to N.M. Beskin [13], the commission took the length of the solar year to be 365 days, 5 hours, 49 minutes and 16 seconds; this value was given in the *Alfonsine tables*, a compendium of astronomical data put together in the thirteenth century for King Alfonso X of Castile. Using this figure they deduced that the Julian year was too long by 10 minutes and 44 seconds, and so would be one day late every 134 years. Therefore, the committee reasoned that one leap year should be omitted every 134 years, which is about three every 400 years.

So, regrettably, we are forced to admit that mathematical hindsight does not entail historical accuracy. Nonetheless, we can argue that the theory of continued fractions is what makes a particular approximation worthwhile, regardless of the methods actually used to obtain it. Even though the astronomers and mathematicians appointed by Gregory did not use continued fractions to obtain their results, they *could not have found* a good calendar which was entirely inconsistent with this topic.

4.4.2 How many semitones should there be in an octave?

In musical theory, the interval of an *octave* contains twelve *semitones*. Musically inclined mathematicians (or mathematically talented musicians) may have wondered if there is anything special about the number twelve. Could one work with a musical system of, say, eleven, thirteen or forty–one semitones to the octave? In this section we shall use continued fractions to see that there are very good reasons for having twelve notes in an octave. For readers who may be unfamiliar with basic musical terminology, a brief summary is given in appendix 4 at the end of this chapter.

There are coherent acoustical reasons for asserting that a combination of two musical notes at different pitches will be pleasing to the ear if the *ratio* of their frequencies is a "simple" rational number. The simplest ratios are $\frac{2}{1}$ and $\frac{3}{2}$; in musical terminology these correspond to the intervals of the *octave* and the *(perfect) fifth* respectively. Suppose that we take a fixed note as the basis of a tonal system, and build upon this foundation two sequences of intervals, one consisting of fifths and the other of octaves. In order to obtain a coherent system of finitely many notes rather than an infinite mess, we require these two sequences to meet again at some point. Suppose, then, that p perfect fifths exactly equal q octaves; in terms of frequencies, we have

$$\left(\frac{3}{2}\right)^p = 2^q .$$

Unfortunately, as is easily proved, this equation has no solutions in integers except for $p = q = 0$, which is musically trivial. So we shall once again employ continued fractions to find the best possible approximate solutions. Rewriting the above equation to find the desired (but unachievable) value of p/q, and then computing its continued fraction, we obtain

$$\frac{p}{q} \approx \frac{\log 2}{\log \frac{3}{2}} = 1 + \frac{1}{1+} \ \frac{1}{2+} \ \frac{1}{2+} \ \frac{1}{3+} \ \frac{1}{1+} \ \frac{1}{5+} \ \frac{1}{2+} \ \frac{1}{23+} \ \frac{1}{2+} \ \cdots ,$$

whose first few convergents are

$$\frac{1}{1}, \ \frac{2}{1}, \ \frac{5}{3}, \ \frac{12}{7}, \ \frac{41}{24}, \ \frac{53}{31} .$$

The first two of these suggest that we should take a fifth, or two fifths, to be the same as an octave: both of these approximations are far too crude to give satisfactory musical results. The third convergent suggests that we approximate five fifths by three octaves: that is, the note five fifths above our fundamental pitch should be replaced by that three octaves above the fundamental. Choosing an arbitrary fundamental for purposes of illustration, our tonal system would then consist of just five notes, where notes differing by an octave are regarded as "the same". These notes, shown in figure 4.1, comprise a *pentatonic* system, and can be used to perform quite satisfying, if perhaps

Figure 4.1 Notes of a pentatonic scale.

Figure 4.2 Notes of the chromatic scale.

extremely simple, music. Indeed, folk music from many parts of the world is found to be based on pentatonic scales.

If we want a system containing more notes, and therefore offering wider musical possibilities, we could turn to the next convergent and take twelve fifths to equal seven octaves. Our (modified) sequence of fifths could look like that displayed in figure 4.2. Transposing and reordering these notes gives the twelve–note chromatic scale which has been the basis of most Western music for many centuries.

We see, then, that the approximation of irrational numbers by rationals, employing as its principal tool the convergents to a continued fraction, suggests a reason why there should be twelve "different" notes used in musical composition, or to put it another way, why there should be twelve semitones in an octave rather than eleven or thirteen. If we seek an even better accommodation of fifths and octaves we may look further along the sequence of convergents: the next would give us a 41–note scale, and if we ignore the limitations of human performance we can make the ratio of fifths to octaves as accurate as we wish.

Despite the beauty of the mathematics involved in this problem, we must again beware of using it as a substitute for history. There is no reason to believe that the chromatic scale was designed with continued fractions in mind. On the contrary, it seems clear that our present tonal system was not "designed" at all but simply developed in accordance with the needs of composers and performers. As in the calendar problem, however, we are very likely justified in asserting that this development is a manifestation of properties of continued fractions. The laws of nature hold in spite of our ignorance of them.

4.5   A "COMPUTATIONAL" TEST FOR RATIONALITY

Continued fractions can sometimes be used to give us an idea (though not necessarily a proof) that a certain number, presented as an infinite decimal, may be rational. For example, in connection with Apéry's irrationality proof for $\zeta(3)$ it was conjectured, see [66], that $\zeta(4)$ can be written as a sum

$$\zeta(4) = c \sum_{n=1}^{\infty} \frac{1}{n^4 \binom{2n}{n}},$$

with $c \in \mathbb{Q}$. Since it is known that $\zeta(4) = \pi^4/90$, the claim is, in effect, that

$$c = \pi^4 \Big/ 90 \sum_{n=1}^{\infty} \frac{1}{n^4 \binom{2n}{n}}$$

is rational. Evaluating c to 10 significant figures (which can be done by taking the first 10 terms of the sum) and then calculating its continued fraction gives

$$c \approx 2.117647059 = 2 + \frac{1}{8+} \ \frac{1}{2+} \ \frac{1}{19607842+} \ \cdots \ .$$

Now the partial quotients in the continued fraction of a "sensible" real number generally consist of fairly small integers. In fact it can be shown (see, for example, Khinchin [35], section 16) that for a "randomly chosen" real number, a proportion about

$$\log_2\Big(\frac{a+1}{a} \Big/ \frac{a+2}{a+1}\Big)$$

of the partial quotients should be equal to a given positive integer a; doing the appropriate calculations, we find that about 42% of the partial quotients should be 1, about 17% should be 2, and so on. One suspects, then, that the partial quotient 19607842 is due to numerical inaccuracy, and that the continued fraction should have terminated at the previous partial quotient. Therefore, it seems reasonable to believe that

$$c = 2 + \frac{1}{8+} \ \frac{1}{2} = \frac{36}{17} \ .$$

Obviously this does not constitute a rigorous proof, but in fact, it is possible to prove that c has the value $\frac{36}{17}$, as conjectured.

Another example of this technique arose in the present author's investigations which ultimately led to the paper [6]. I sought to evaluate **generalised continued fractions** of the form

$$f(s) = \cfrac{1+s}{1 + \cfrac{2+s}{2 + \cfrac{3+s}{3+\cdots}}}$$

for positive integers s. Computing decimal approximations of these expressions for a few values of s, and then writing them as simple continued fractions gave results such as

$$f(4) = 1.863013699$$
$$= 1 + \cfrac{1}{1+} \cfrac{1}{6+} \cfrac{1}{3+} \cfrac{1}{3+} \cfrac{1}{507356+} \cdots$$
$$f(5) = 2.085828343$$
$$= 2 + \cfrac{1}{2+} \cfrac{1}{11+} \cfrac{1}{1+} \cfrac{1}{1+} \cfrac{1}{1+} \cfrac{1}{6+} \cfrac{1}{1+} \cfrac{1}{1+} \cfrac{1}{13733+} \cdots.$$

As explained above, these calculations led to the conjectures

$$f(4) = 1 + \cfrac{1}{1+} \cfrac{1}{6+} \cfrac{1}{3+} \cfrac{1}{3} = \frac{136}{73}$$

and

$$f(5) = 2 + \cfrac{1}{2+} \cfrac{1}{11+} \cfrac{1}{1+} \cfrac{1}{1+} \cfrac{1}{1+} \cfrac{1}{6+} \cfrac{1}{1+} \cfrac{1}{1} = \frac{1045}{501}.$$

Moreover, similar outcomes were observed for all tested values of s, and so it was conjectured that $f(s)$ is always rational when s is a positive integer. After a good deal of further work it was found possible to prove this assertion, and to confirm the specific values conjectured above. An informal account of the investigation is given in [5].

4.6 FURTHER APPROXIMATION PROPERTIES OF CONVER-GENTS

We know that for every irrational α the inequality

$$\left| \alpha - \frac{p}{q} \right| < \frac{1}{q^2}$$

has infinitely many solutions, and that for certain α the right–hand side can be decreased by substituting for q^2 a higher power q^s. Indeed, if α is a Liouville number then we can choose arbitrarily large s and still find infinitely many solutions; on the other hand, we know from Liouville's Theorem, page 49, that if α is a quadratic irrational then the exponent 2 cannot be increased at all.

Thus, if we want a result which is true for all α, we cannot decrease the right–hand side of the above inequality by increasing s. Perhaps, however, we could replace the 1 in the numerator by something smaller. Indeed, we can, for the comment following Theorem 4.13 shows that 1 can be replaced by $\frac{1}{2}$. Can we do even better than this?

Recall that convergents give "the best" approximations to a real number α, and that the approximations are especially good when the next partial quotient is large. Consider what happens if the "next partial quotient" of α is *never* large. An extreme example of such a number is

$$\alpha = 1 + \frac{1}{1+} \, \frac{1}{1+} \, \frac{1}{1+} \cdots = \frac{1+\sqrt{5}}{2} \, .$$

In this case it is plain that every complete quotient α_k is equal to α. Moreover, if we calculate the first few convergents to α it is very easy to conjecture, and equally easy to prove by induction, that $q_k = p_{k-1}$ for $k \geq -1$. (It is also easy to show, though unimportant at present, that the numerators p_k and denominators q_k are just the Fibonacci numbers.) Therefore, from equation (4.5), we have

$$\left| \alpha - \frac{p_k}{q_k} \right| = \frac{1}{(\alpha q_k + q_{k-1})q_k} = \frac{1}{(\alpha + q_{k-1}/q_k)q_k^2} \approx \frac{1}{(\alpha + \alpha^{-1})q_k^2} = \frac{1}{\sqrt{5}\,q_k^2} \, .$$

Since we do not expect that any real irrational number will have *worse* rational approximations than α, the following result is plausible.

Theorem 4.14. (Hurwitz). *Let α be a real irrational number. Then the inequality*

$$\left| \alpha - \frac{p}{q} \right| < \frac{1}{\sqrt{5}\,q^2}$$

has infinitely many rational solutions p/q.

Adolf Hurwitz
(1859–1919)

Proof. Write

$$r_k = \frac{q_{k-1}}{q_k}$$

for $k \geq 0$. The idea of the proof is to show that if the required inequality fails for three consecutive convergents p_k/q_k, p_{k+1}/q_{k+1} and p_{k+2}/q_{k+2} to α, then the two consecutive quotients r_{k+1} and r_{k+2} are both strictly greater than $\frac{1}{2}(\sqrt{5}-1)$, and a contradiction follows. To begin the proof, we use the equality in (4.5) to see that

$$\left| \alpha - \frac{p_k}{q_k} \right| < \frac{1}{\sqrt{5}\,q_k^2} \quad \text{if and only if} \quad \alpha_{k+1} + r_k > \sqrt{5} \, . \tag{4.8}$$

Suppose that in fact

$$\alpha_{k+1} + r_k \leq \sqrt{5} \quad \text{and} \quad \alpha_{k+2} + r_{k+1} \leq \sqrt{5} \, , \tag{4.9}$$

and observe that we have

$$\alpha_{k+1} = a_{k+1} + \frac{1}{\alpha_{k+2}} \quad \text{and} \quad \frac{1}{r_{k+1}} = a_{k+1} + r_k \, , \tag{4.10}$$

where the second identity follows immediately from the defining recursion for q_{k+1}. Eliminating a_{k+1} we obtain

$$\frac{1}{\alpha_{k+2}} + \frac{1}{r_{k+1}} = \alpha_{k+1} + r_k ,$$

and by using the assumed inequalities (4.9), we have

$$\frac{1}{\sqrt{5} - r_{k+1}} + \frac{1}{r_{k+1}} \leq \sqrt{5} .$$

This inequality is easily simplified to give

$$r_{k+1}^2 - \sqrt{5}\, r_{k+1} + 1 \leq 0 ,$$

and since the roots of the quadratic $x^2 - \sqrt{5}\, x + 1$ are $\frac{1}{2}(\sqrt{5} \pm 1)$ we obtain $r_{k+1} \geq \frac{1}{2}(\sqrt{5} - 1)$. But r_{k+1} is rational, so equality cannot hold and we have

$$r_{k+1} > \frac{\sqrt{5} - 1}{2} .$$

If we now suppose further that $\alpha_{k+3} + r_{k+2} \leq \sqrt{5}$, then exactly the same argument gives

$$r_{k+2} > \frac{\sqrt{5} - 1}{2} ;$$

but using the second identity from (4.10), with k replaced by $k + 1$, we have

$$1 = (a_{k+2} + r_{k+1})r_{k+2} > \left(1 + \frac{\sqrt{5} - 1}{2}\right)\frac{\sqrt{5} - 1}{2} = 1 ,$$

which is absurd. To sum up, we have shown that a contradiction follows from the assumption that the inequalities (4.8) are false for three consecutive integers k; consequently, the required inequality is true for at least one in every three convergents p_k/q_k, and the theorem is proved.

Theorem 4.15. Hurwitz' constant is best possible. *The constant $\sqrt{5}$ in Hurwitz' Theorem is the best possible. That is, if A is a constant greater than $\sqrt{5}$, then there exist irrational numbers α such that the inequality*

$$\left|\alpha - \frac{p}{q}\right| < \frac{1}{Aq^2} \tag{4.11}$$

has only finitely many rational solutions.

Proof. We shall show that $\alpha = \frac{1}{2}(1 + \sqrt{5})$ is such a number – no surprise, in view of the comments introducing Hurwitz' Theorem. Let $A > \sqrt{5}$. For any solution p/q of (4.11), we have

$$\left|\alpha - \frac{p}{q}\right| < \frac{1}{\sqrt{5}\, q^2} < \frac{1}{2q^2} ,$$

and so by a previous result p/q must be a convergent p_k/q_k to α. Using a result asserted on page 88 (proof: see exercise 4.2), we have

$$\left| \alpha - \frac{p}{q} \right| = \frac{1}{(\alpha_{k+1}q_k + q_{k-1})q_k} = \frac{1}{(\alpha + q_{k-1}/p_{k-1})q^2} ,$$

and therefore

$$A < \alpha + \frac{q_{k-1}}{p_{k-1}} .$$

It follows that (4.11) has only a finite number of solutions. For otherwise this inequality would hold for arbitrarily large k and we would find

$$A \leq \lim_{k \to \infty} \left(\alpha + \frac{q_{k-1}}{p_{k-1}} \right) = \alpha + \frac{1}{\alpha} = \sqrt{5} ,$$

contrary to assumption. This completes the proof.

Comments.

- An alternative proof of this result has echoes of Liouville's proof in Chapter 3. Let

$$\alpha = \frac{1 + \sqrt{5}}{2} .$$

Then the minimal polynomial of α is $f(x) = x^2 - x - 1$, and for any rational number p/q we have by the Mean Value Theorem

$$\frac{f(\alpha) - f(p/q)}{\alpha - p/q} = f'(\gamma) = 2\gamma - 1$$

for some γ between α and p/q. However, $f(\alpha) = 0$, and

$$\left| f\left(\frac{p}{q}\right) \right| = \left| \frac{p^2 - pq - q^2}{q^2} \right| \geq \frac{1}{q^2} ,$$

and

$$|2\gamma - 1| = |(2\alpha - 1) + 2(\gamma - \alpha)| \leq \sqrt{5} + 2 \left| \alpha - \frac{p}{q} \right| ;$$

putting all this information together yields

$$\frac{1}{q^2} \leq \left(\sqrt{5} + 2 \left| \alpha - \frac{p}{q} \right| \right) \left| \alpha - \frac{p}{q} \right| .$$

All of this is true for any p/q; if p/q is a solution of (4.11), then

$$A < \sqrt{5} + 2 \left| \alpha - \frac{p}{q} \right| < \sqrt{5} + \frac{2}{Aq^2} .$$

Now suppose that $A > \sqrt{5}$. Then

$$q^2 < \frac{2}{A(A - \sqrt{5})} ,$$

so there are only finitely many possibilities for q, and (4.11) can have only finitely many rational solutions.

- In the proof of Theorem 4.15 we considered the number $\alpha = \frac{1}{2}(1 + \sqrt{5})$ and made use of the fact that every complete quotient of α is equal to α itself. In fact, it would have been sufficient to consider a number β having infinitely many complete quotients equal to α. A little thought shows that this is true if and only if

$$\beta = b_0 + \cfrac{1}{b_1 +} \; \cfrac{1}{\ldots +} \; \cfrac{1}{b_n +} \; \cfrac{1}{1 +} \; \cfrac{1}{1 +} \; \cfrac{1}{1 +} \; \cdots$$

for some integers $b_0, b_1, \ldots, b_n$; in other words, if the partial quotients of β are all 1 from some point on. It can be shown that if α is such a number then for $A > \sqrt{5}$ the inequality (4.11) has only a finite number of solutions. On the other hand, if we exclude from consideration all such α then the result can be improved, and we find that

$$\left| \alpha - \frac{p}{q} \right| < \frac{1}{\sqrt{8}\, q^2}$$

for infinitely many rationals p/q. If even more α are ignored, then the constant $\sqrt{8}$ can be increased still further. In fact, Hurwitz' Theorem is the first of a series of results which show that if α has a continued fraction *not* in a certain finite number of categories, then there is a constant C such that

$$\left| \alpha - \frac{p}{q} \right| < \frac{1}{C q^2}$$

has infinitely many solutions; the constants C can be explicitly identified, and can be shown to be best possible, in the sense of Theorem 4.15. These constants make up a (countable) set

$$\left\{ \sqrt{9 - \frac{4}{m^2}} \ \middle|\ m = 1, 2, 5, 13, 29, 34, \ldots \right\}$$

which forms part of what is known as the **Lagrange spectrum**. More information (particularly on where the sequence $1, 2, 5, 13, 29, 34, \ldots$ comes from) may be found in Bombieri [16].

We can make use of properties of continued fractions to show, as asserted in Chapter 3, that there exist transcendental numbers which are not Liouville numbers.

Theorem 4.16. *Any Liouville number has a continued fraction with unbounded partial quotients.*

Proof. Suppose that α has partial quotients $a_k < A$ for $k \geq 1$. Now if α is rational it is certainly not a Liouville number; suppose that α is irrational. If α is approximable to order $s > 2$, then there is a constant c such that

$$0 < \left| \alpha - \frac{p}{q} \right| < \frac{c}{q^s} \tag{4.12}$$

for rationals p/q with arbitrarily large q. Choose q sufficiently large that $q^{s-2} > (A+2)c$. Then certainly $q^{s-2} > 2c$, we have

$$0 < \left| \alpha - \frac{p}{q} \right| < \frac{1}{2q^2} \,,$$

and by a previous result p/q is one of the convergents to α. For $p/q = p_k/q_k$ we have

$$\left| \alpha - \frac{p}{q} \right| = \frac{1}{(\alpha_{k+1}q_k + q_{k-1})q_k} > \frac{1}{(a_{k+1}+2)q_k^2} > \frac{1}{(A+2)q^2} \,,$$

and since $q^{s-2} > (A+2)c$ this contradicts (4.12). Hence α is not approximable to order greater than 2, which proves considerably more than we claimed.

Corollary 4.17. *Not all transcendental numbers are Liouville numbers.*

Proof. Consider all possible continued fractions

$$a_0 + \cfrac{1}{a_1 +} \cfrac{1}{a_2 +} \cfrac{1}{a_3 + \cdots} \,, \tag{4.13}$$

where each a_k is either 1 or 2. It is a standard result of set theory that the set of all such continued fractions is uncountable; but the set of algebraic numbers is countable. So there exist (uncountably many!) continued fractions of the form (4.13) which are transcendental. But by the above theorem, none of these continued fractions is a Liouville number.

From these results we know that certain transcendental numbers have continued fractions with unbounded partial quotients while others have continued fractions with bounded partial quotients (see exercise 4.10 for a specific example of the latter). What happens if we ask similar questions regarding algebraic numbers? We already know that the continued fraction expansions for rationals and for quadratic irrationals are finite and periodic respectively, and hence clearly have bounded partial quotients, so we need only consider algebraic numbers of degree higher than 2.

- Do there exist algebraic numbers of degree $n \geq 3$ whose continued fractions have bounded partial quotients?

- Do there exist algebraic numbers of degree $n \geq 3$ whose continued fractions have unbounded partial quotients?

Clearly one, perhaps both, of these problems must have an affirmative answer; unfortunately, each remains unsolved! There is not a single algebraic number of degree 3 or more whose complete continued fraction is known (though most experts in the field believe that such a number will *always* have unbounded partial quotients). Indeed, there appears to be very little of any great generality that can be said about continued fractions for algebraic numbers of degree 3 or more, or for transcendental numbers. We can, however, give some useful methods of computation, as well as some results for specific important numbers.

4.7 COMPUTING THE CONTINUED FRACTION OF AN ALGE-BRAIC IRRATIONAL

Let α be a root of a known irreducible polynomial f with degree $n \geq 2$ and integral coefficients. We shall assume that $\alpha > 1$, and that f has no other roots $\beta > 1$. In this case there is a very simple algorithm to find the zeroth partial quotient $a_0 = \lfloor \alpha \rfloor$: calculate $f(1), f(2), f(3), \ldots$ until a change of sign occurs; then a_0 is the last argument before the change of sign. Since α is irrational we have $a_0 < \alpha < a_0 + 1$. The first complete quotient α_1 is defined by $\alpha = a_0 + 1/\alpha_1$; therefore

$$f\left(a_0 + \frac{1}{\alpha_1}\right) = 0 ,$$

and α_1 is a root of the polynomial defined by

$$f_1(z) = z^n f\left(a_0 + \frac{1}{z}\right) .$$

Since, by assumption, f has a unique real root greater than 1, it is not hard to show that f_1 has the same property. For let $\beta = \alpha_1 = 1/(\alpha - a_0)$. Then

$$f_1(\beta) = \beta^n f\left(a_0 + \frac{1}{\beta}\right) = \beta^n f(\alpha) = 0 ,$$

and $\beta > 1$ since $0 < \alpha - a_0 < 1$; so f_1 has a real root greater than 1. Conversely, if β is any such root of f_1, then $a_0 + 1/\beta$ is a root of f and hence $a_0 + 1/\beta = \alpha$. Because f_1 is a polynomial having integral coefficients and a unique real root $\alpha_1 > 1$, the procedure can be iterated to find the sequence of partial quotients of α. Observe that the complete quotients $\alpha = \alpha_0, \alpha_1, \alpha_2, \ldots$ need never be calculated, so we do not have the problem of calculating decimal expansions to many places: all our calculations will be performed in terms of integer arithmetic, and so the process will be free of rounding errors.

Example. We can find the continued fraction for $\sqrt[3]{2}$ by starting with the polynomial $f(z) = f_0(z) = z^3 - 2$. We have

$$f_0(z) = z^3 - 2 , \qquad a_0 = 1 ;$$
$$f_1(z) = -z^3 + 3z^2 + 3z + 1 , \qquad a_1 = 3 ;$$
$$f_2(z) = 10z^3 - 6z^2 - 6z - 1 , \qquad a_2 = 1 ;$$
$$f_3(z) = -3z^3 + 12z^2 + 24z + 10 , \qquad a_3 = 5 ;$$
$$f_4(z) = 55z^3 - 81z^2 - 33z - 3 , \qquad a_4 = 1 ;$$
$$f_5(z) = -62z^3 - 30z^2 + 84z + 55 , \qquad a_5 = 1 ;$$
$$f_6(z) = 47z^3 - 162z^2 - 216z - 62 , \qquad a_6 = 4$$

and so

$$\sqrt[3]{2} = 1 + \frac{1}{3+} \frac{1}{1+} \frac{1}{5+} \frac{1}{1+} \frac{1}{1+} \frac{1}{4+} \cdots .$$

Alternatively, we can find the continued fraction for $\alpha = \sqrt[3]{2}$ by adapting the method used for quadratic irrationals. However, α has degree 3 and we shall therefore, in general, have to find reciprocals of sums of three terms. Writing $\omega = e^{2\pi i/3}$, we have

$$\frac{1}{a + b\sqrt[3]{2} + c\sqrt[3]{4}} = \frac{\left(a + b\omega\sqrt[3]{2} + c\omega^2\sqrt[3]{4}\right)\left(a + b\omega^2\sqrt[3]{2} + c\omega\sqrt[3]{4}\right)}{a^3 + 2b^3 + 4c^3 - 6abc}$$

$$= \frac{a^2 - 2bc}{a^3 + 2b^3 + 4c^3 - 6abc} + \frac{2c^2 - ab}{a^3 + 2b^3 + 4c^3 - 6abc}\sqrt[3]{2}$$

$$+ \frac{b^2 - ac}{a^3 + 2b^3 + 4c^3 - 6abc}\sqrt[3]{4} ,$$

and clearly the algebra is going to be significantly harder than it is in the quadratic case. The polynomial method involves much less work.

Comment. In the above discussion we assumed that f_0 has no real root $\beta > 1$ except for $\beta = \alpha$. If this is not true we may apply the same algorithm anyway, and obtain an appropriate sequence of functions f_k. The only possible additional difficulty lies in determining the partial quotients $a_k = \lfloor \alpha_k \rfloor$: it may be that f_k has more than one root greater than 1, and we need to ensure that we choose the correct one as α_k. However, it can be shown that the assumed condition must be true, if not for f_0, then after a certain number of iterations, and once this happens the algorithm proceeds with little difficulty.

Example. Find the first few terms of the continued fraction of α, the smallest positive root of

$$f(z) = 2z^3 - 20z^2 + 62z - 61 .$$

Solution. Being careful not to miss a root, we find that f has two roots between 2 and 3, and one between 5 and 6. So $a_0 = 2$ and

$$f_1(z) = z^3 f\left(2 + \frac{1}{z}\right) = -z^3 + 6z^2 - 8z + 2 .$$

Now f_1 has roots between 0 and 1, which is of no interest to us, between 1 and 2, and between 4 and 5. The smallest root of f corresponds to the largest root of f_1; therefore $a_1 = 4$ and

$$f_2(z) = z^3 f_1\left(4 + \frac{1}{z}\right) = 2z^3 - 8z^2 - 6z - 1 .$$

We find that f_2 has only one root greater than 1 and so the procedure is standard from now on; we obtain

$$\alpha = 2 + \frac{1}{4+} \ \frac{1}{4+} \ \frac{1}{1+} \ \frac{1}{1+} \ \frac{1}{1+} \ \frac{1}{149+} \ \frac{1}{3+} \ \cdots .$$

Another point worth noting is that the procedure which locates a root of a polynomial by searching for a sign change will fail for a double root. In this case, however, we chose our original polynomial f to be irreducible, and it follows from exercise 3.11 that f cannot have a double root.

4.8   THE CONTINUED FRACTION OF e

We shall determine the continued fractions for a class of numbers related to the exponential constant e. To do so, we first consider the functions defined by the infinite series

$$f(c; z) = \sum_{k=0}^{\infty} \frac{1}{c(c+1) \cdots (c+k-1)} \frac{z^k}{k!} \; ;$$

here the parameter c is any real number except $0, -1, -2, \ldots$, and it is easy to show that the series converges for all z. To simplify the notation we write $c^{(k)} = c(c+1) \cdots (c+k-1)$, with the understanding that $c^{(0)} = 1$; thus

$$f(c; z) = \sum_{k=0}^{\infty} \frac{1}{c^{(k)}} \frac{z^k}{k!} \; .$$

The expression $c^{(k)}$ is referred to as "c rising factorial k". It satisfies the two important recurrences

$$c^{(k+1)} = c^{(k)}(c+k) = c(c+1)^{(k)} \; , \tag{4.14}$$

both of which are instances of the more general relation $c^{(k+m)} = c^{(k)}(c+k)^{(m)}$.

Lemma 4.18. *Let c be a positive real number, z a non–zero real number and k a non–negative integer; then*

$$\frac{c}{z} \frac{f(c; z^2)}{f(c+1; z^2)} = \left[\frac{c}{z}, \frac{c+1}{z}, \ldots, \frac{c+k-1}{z}, \frac{c+k}{z} \frac{f(c+k; z^2)}{f(c+k+1; z^2)} \right] \; .$$

Proof. First, observe that under the stated conditions $f(c+k+1; z^2)$ is given by a series of positive terms, so it does not vanish and the last term in the continued fraction makes sense. From (4.14) the rising factorials satisfy

$$\frac{1}{c^{(k)}} = \frac{c+k}{c^{(k+1)}} = \frac{1}{(c+1)^{(k)}} + \frac{k}{c^{(k+1)}} \; ,$$

and hence

$$f(c; z) = \sum_{k=0}^{\infty} \frac{1}{(c+1)^{(k)}} \frac{z^k}{k!} + \sum_{k=0}^{\infty} \frac{k}{c^{(k+1)}} \frac{z^k}{k!} \; .$$

The first series on the right–hand side is evidently $f(c+1; z)$. The second may be written

$$\sum_{k=1}^{\infty} \frac{1}{c^{(k+1)}} \frac{z^k}{(k-1)!} = \sum_{k=0}^{\infty} \frac{1}{c^{(k+2)}} \frac{z^{k+1}}{k!} = \frac{z}{c(c+1)} \sum_{k=0}^{\infty} \frac{1}{(c+2)^{(k)}} \frac{z^k}{k!} \; ,$$

and we have the second–order recurrence

$$f(c; z) = f(c+1; z) + \frac{z}{c(c+1)} f(c+2; z) \; .$$

Rearranging this equation and replacing z by z^2 gives an identity which looks something like the beginning of a continued fraction,

$$f(c; z^2)/f(c+1; z^2) = 1 + \frac{z^2/c(c+1)}{f(c+1; z^2)/f(c+2; z^2)} ;$$

however, we need to convert this into a form in which the numerator $z^2/c(c+1)$ is replaced by 1. So, multiply the quotient $f(c; z^2)/f(c+1; z^2)$ by a factor A_c to give

$$A_c f(c; z^2)/f(c+1; z^2) = A_c + \frac{A_c z^2/c(c+1)}{f(c+1; z^2)/f(c+2; z^2)}$$

$$= A_c + \frac{A_c A_{c+1} z^2/c(c+1)}{A_{c+1} f(c+1; z^2)/f(c+2; z^2)} . \quad (4.15)$$

We want $A_c A_{c+1} z^2/c(c+1) = 1$, that is, $A_c A_{c+1} = c(c+1)/z^2$, and it will suffice to take $A_c = c/z$. Making this choice and substituting into (4.15), we have

$$\frac{c}{z} \frac{f(c; z^2)}{f(c+1; z^2)} = \frac{c}{z} + \frac{1}{\dfrac{c+1}{z} \dfrac{f(c+1; z^2)}{f(c+2; z^2)}} .$$

But now the complete quotient on the right–hand side can be treated in the same way, and the whole process repeated k times, to yield

$$\frac{c}{z} \frac{f(c; z^2)}{f(c+1; z^2)} = \left[\frac{c}{z}, \frac{c+1}{z}, \dots, \frac{c+k-1}{z}, \frac{c+k}{z} \frac{f(c+k; z^2)}{f(c+k+1; z^2)} \right]$$

as claimed.

Theorem 4.19. *Let c and z be real numbers with $c > 0$ and $z \neq 0$, such that $(c+k)/z$ is a positive integer for all $k \geq 0$. Then*

$$\frac{c}{z} \frac{f(c; z^2)}{f(c+1; z^2)} = \left[\frac{c}{z}, \frac{c+1}{z}, \frac{c+2}{z}, \dots \right] .$$

Proof. Using the previous lemma and the uniqueness result Lemma 4.7, all we need show is that

$$\frac{c+k}{z} \frac{f(c+k; z^2)}{f(c+k+1; z^2)} > 1 \quad (4.16)$$

for $k \geq 0$. But we have

$$(c+k+1)^{(k)} = (c+k+1)(c+k+2) \cdots (c+2k)$$
$$> (c+k)(c+k+1) \cdots (c+2k-1)$$
$$= (c+k)^{(k)} ,$$

and noting that z^2 is a positive real number, it follows that

$$f(c+k; z^2) = \sum_{k=0}^{\infty} \frac{1}{(c+k)^{(k)}} \frac{z^{2k}}{k!} > \sum_{k=0}^{\infty} \frac{1}{(c+k+1)^{(k)}} \frac{z^{2k}}{k!} = f(c+k+1; z^2) \ .$$

Since by assumption $(c+k)/z \geq 1$, the inequality (4.16) holds and the result is proved.

Finally we must find c and z such that the conditions in the above result hold, and also such that the series representing $f(c; z^2)$ and $f(c+1; z^2)$ can be summed without excessive difficulty. We need c/z and $1/z$ to be positive integers, say $c/z = p$ and $1/z = q$; then

$$p \frac{f\left(\frac{p}{q}; \frac{1}{q^2}\right)}{f\left(\frac{p}{q}+1; \frac{1}{q^2}\right)} = [\, p,\, p+q,\, p+2q,\, p+3q, \ldots\,] \ .$$

Moreover, if $k \geq 0$, we have

$$\left(\tfrac{1}{2}\right)^{(k)} = \left(\tfrac{1}{2}\right)\left(\tfrac{3}{2}\right) \cdots \left(k - \tfrac{1}{2}\right) = \frac{\left(\frac{1}{2}\right)(1)\left(\frac{3}{2}\right)(2) \cdots \left(k - \frac{1}{2}\right)(k)}{k!} = \frac{(2k)!}{2^{2k} k!}$$

and so

$$f\left(\tfrac{1}{2}; z^2\right) = \sum_{k=0}^{\infty} \frac{2^{2k} k!}{(2k)!} \frac{z^{2k}}{k!} = \sum_{k=0}^{\infty} \frac{(2z)^{2k}}{(2k)!} = \cosh 2z \ .$$

Similarly $f\left(\tfrac{3}{2}; z^2\right) = (2z)^{-1} \sinh 2z$; taking $q = 2p$ and combining these results,

$$\frac{\cosh \frac{1}{p}}{\sinh \frac{1}{p}} = p \frac{f\left(\frac{1}{2}; \frac{1}{4p^2}\right)}{f\left(\frac{3}{2}; \frac{1}{4p^2}\right)} \ .$$

We have proved the following result.

Theorem 4.20. *For any positive integer p, we have*

$$\coth \frac{1}{p} = p + \frac{1}{3p+} \ \frac{1}{5p+} \ \frac{1}{7p+} \ \cdots \ .$$

Corollary 4.21. *The numbers*

$$\coth \frac{1}{p} \ , \quad \frac{e+1}{e-1} \quad and \quad e \ ,$$

where in the first p is a positive integer, are neither rational numbers nor quadratic irrationalities.

Proof. The first claim is true because $\coth 1/p$ has an infinite non–periodic continued fraction. The second number given is in fact $\coth \frac{1}{2}$. If e were rational or a quadratic irrational, then $(e+1)/(e-1)$ would be too.

Comment. The function $f(c; z)$ employed in the above proof is closely related to the **hypergeometric function**, defined by the series

$$F(a, b; c; z) = \sum_{k=0}^{\infty} \frac{a(a+1)\cdots(a+k-1)\, b(b+1)\cdots(b+k-1)}{c(c+1)\cdots(c+k-1)} \frac{z^k}{k!}$$

for $|z| < 1$. Here a, b and c are real parameters with $c \neq 0, -1, -2, \ldots$. The hypergeometric function is a solution of the differential equation

$$z(1-z)y'' + (c - (a+b+1)z)y' - aby = 0 .$$

A large number of important functions are special cases of the hypergeometric function. For example,

$$F(1, 1; 1; z) = \sum_{k=0}^{\infty} z^k = \frac{1}{1-z} ,$$

$$F(-n, 1; 1; -z) = \sum_{k=0}^{\infty} \frac{n(n-1)\cdots(n-k+1)}{k!} z^k = (1+z)^n \quad \text{for } n \in \mathbb{N} ,$$

$$zF(1, 1; 2; z) = \sum_{k=0}^{\infty} \frac{z^{k+1}}{k+1} = -\log(1-z) ;$$

while by taking $c = \frac{1}{2}$ in the recurrence (4.14) we find $(\frac{3}{2})^{(k)}/(\frac{1}{2})^{(k)} = 2k+1$ and hence

$$zF\left(\tfrac{1}{2}, 1; \tfrac{3}{2}; -z^2\right) = \sum_{k=0}^{\infty} \frac{(\frac{1}{2})^{(k)}}{(\frac{3}{2})^{(k)}} (-1)^k z^{2k+1} = \sum_{k=0}^{\infty} (-1)^k \frac{z^{2k+1}}{2k+1} = \tan^{-1} z .$$

The functions cosh and sinh cannot be obtained by making a simple substitution of particular values for a, b and c; however, we have

$$z \lim_{a\to\infty} \lim_{b\to\infty} F(a, b; \tfrac{3}{2}; z^2/4ab) = \sum_{k=0}^{\infty} \left(\lim_{a\to\infty} \frac{a^{(k)}}{a^k} \right) \left(\lim_{b\to\infty} \frac{b^{(k)}}{b^k} \right) \frac{2^{2k} k!}{(2k+1)!} \frac{z^{2k+1}}{2^{2k} k!}$$

$$= \sum_{k=0}^{\infty} \frac{z^{2k+1}}{(2k+1)!}$$

$$= \sinh z ,$$

it being possible to justify the interchange of sum and limits.

By using the continued fractions found above we may compute the continued fraction for e itself.

Theorem 4.22. *The continued fraction of e is*

$$e = 2 + \cfrac{1}{1+}\ \cfrac{1}{2+}\ \cfrac{1}{1+}\ \cfrac{1}{1+}\ \cfrac{1}{4+}\ \cfrac{1}{1+}\ \cfrac{1}{1+}\ \cfrac{1}{6+}\ \cfrac{1}{1+}\ \cdots .$$

Proof. Let the convergents of the continued fractions $\alpha = [2, 6, 10, 14, \ldots]$ and $\beta = [1, 1, 2, 1, 1, 4, 1, 1, 6, \ldots]$ be p_k/q_k and r_k/s_k respectively. The kth partial quotient of the former is $4k + 2$, and so we have

$$p_k = (4k + 2)p_{k-1} + p_{k-2}, \quad q_k = (4k + 2)q_{k-1} + q_{k-2}.$$

The partial quotients of the second continued fraction are

$$a_m = \begin{cases} 2k & \text{if } m = 3k - 1 \\ 1 & \text{otherwise,} \end{cases}$$

and so for any $k \geq 1$ we can write

$$r_{3k+1} = r_{3k} + r_{3k-1}$$
$$r_{3k} = r_{3k-1} + r_{3k-2}$$
$$r_{3k-1} = 2k\, r_{3k-2} + r_{3k-3}$$
$$r_{3k-2} = r_{3k-3} + r_{3k-4}$$
$$r_{3k-3} = r_{3k-4} + r_{3k-5}.$$

Eliminating $r_{3k}, r_{3k-1}, r_{3k-3}$ and r_{3k-4} from these equations gives a recurrence involving r_{3k+1}, r_{3k-2} and r_{3k-5}; the same recurrence is satisfied by the corresponding denominators, and we have

$$r_{3k+1} = (4k + 2)r_{3k-2} + r_{3k-5}, \quad s_{3k+1} = (4k + 2)s_{3k-2} + s_{3k-5}.$$

Since p_k and q_k satisfy the same relations, a little attention to initial values and an easy induction shows that

$$r_{3k+1} = 2q_k \quad \text{and} \quad s_{3k+1} = p_k - q_k$$

for all $k \geq -1$. From Theorem 4.20 we have $\alpha = \coth \frac{1}{2} = (e + 1)/(e - 1)$; therefore

$$\beta = \lim_{k \to \infty} \frac{r_k}{s_k} = \lim_{k \to \infty} \frac{r_{3k+1}}{s_{3k+1}} = \lim_{k \to \infty} \frac{2q_k}{p_k - q_k} = \frac{2}{\alpha - 1} = e - 1,$$

from which we obtain the continued fraction for e.

The continued fractions for e and related numbers were first determined by Euler ([25]; English translation in [70]). He begins with a certain differential equation, and subsequent calculations involve an expression which in the notation used here is $f(1 + \frac{1}{n}; \frac{z}{n})$. Earlier in the same paper, he states that a number is rational if and only if its simple continued fraction terminates; from which it may be deduced that e is irrational. Curiously, Euler does not explicitly draw this conclusion; perhaps he felt that it was too obvious to be worth writing down. An interesting historical discussion of [25] may be found in [21]. A different method of calculating the continued fraction of e was published by Hermite [31] in 1873; it involves an integral very similar to the one we have used in Chapter 2 to prove the irrationality of e^r. Accessible expositions of Hermite's method are given in Olds [47] and Cohn [20].

EXERCISES

4.1 (a) Find the continued fraction of $\dfrac{1229}{321}$.

(b) Find the continued fraction of $\sqrt{a^2 + a + \frac{1}{2}}$, where a is a positive integer.

(c) Evaluate the eventually periodic continued fraction

$$1 + \cfrac{1}{2+} \cfrac{1}{1+} \cfrac{1}{1+} \cfrac{1}{3+} \cfrac{1}{1+} \cfrac{1}{1+} \cfrac{1}{3+} \cfrac{1}{1+} \cfrac{1}{1+} \cfrac{1}{3+} \cdots .$$

4.2 Consider the continued fraction

$$\alpha = a_0 + \cfrac{1}{a_1 +} \cdots \cfrac{1}{a_n} ,$$

where all a_k (including a_0) are positive integers. As usual, write p_k/q_k for the convergents to α.

(a) Find the continued fraction of p_n/p_{n-1}.

(b) Show that the sequence of partial quotients is palindromic (that is, $a_0 = a_n$, $a_1 = a_{n-1}$ and so on) if and only if $p_{n-1} = q_n$.

4.3 Simplify $p_{k+2}q_k - p_k q_{k+2}$ and $p_{k+3}q_k - p_k q_{k+3}$. Generalise.

4.4 Prove *Euler's rule* for computing the convergents of a continued fraction. Write down the product $a_0 a_1 \cdots a_n$ and all products which can be obtained by deleting any number of pairs $a_k a_{k+1}$ of adjacent factors from this product; if n is odd, one of the products obtained contains no factors and is taken to be 1. Let p_n be the sum of all these products. Find q_n by applying a similar process to the product $a_1 a_2 \cdots a_n$. Then

$$\frac{p_n}{q_n} = a_0 + \cfrac{1}{a_1 +} \cfrac{1}{\dots +} \cfrac{1}{a_n} .$$

4.5 Let a_0 be an integer and $a_1, a_2, \ldots$ positive integers. Evaluate the matrix product

$$\begin{pmatrix} a_0 & 1 \\ 1 & 0 \end{pmatrix} \begin{pmatrix} a_1 & 1 \\ 1 & 0 \end{pmatrix} \cdots \begin{pmatrix} a_k & 1 \\ 1 & 0 \end{pmatrix} ,$$

and hence give a proof, different from that in Lemma 4.1, of the relation

$$p_{k-1}q_k - p_k q_{k-1} = (-1)^k .$$

4.6 Evaluate

$$a_0 + \cfrac{1}{a_1 +} \cdots \cfrac{1}{a_k +} \cfrac{1}{a_k +} \cdots \cfrac{1}{a_1 +} \cfrac{1}{a_0}$$

in terms of the partial numerators and denominators of $[a_0, a_1, \ldots, a_k]$. Ensure that your answer is given as a fraction in lowest terms.

4.7 Consider the periodic continued fraction

$$\alpha = a_0 + \cfrac{1}{a_1 +} \cdots \cfrac{1}{a_{n-1} +} \cfrac{1}{2a_0 +} \cfrac{1}{a_1 +} \cdots \cfrac{1}{a_{n-1} +} \cfrac{1}{2a_0 +} \cdots ,$$

where the partial quotients $a_0, a_1, \ldots, a_{n-1}$ are positive integers and the sequence $a_1, \ldots, a_{n-1}$ is palindromic. Prove that α^2 is rational.

4.8 For any quadratic irrational $\xi = x + \sqrt{y}$ with $x, y \in \mathbb{Q}$ and $\sqrt{y} \notin \mathbb{Q}$, we write ξ^* for the conjugate of ξ, in the sense of Chapter 3: that is, $\xi^* = x - \sqrt{y}$. The aim of the present exercise is to prove that a quadratic irrational α has a *purely periodic* continued fraction

$$\alpha = a_0 + \cfrac{1}{a_1 +} \cdots \cfrac{1}{a_{n-1} +} \cfrac{1}{a_0 +} \cfrac{1}{a_1 +} \cdots \cfrac{1}{a_{n-1} +} \cfrac{1}{a_0 +} \cdots .$$

with all a_k positive integers, if and only if $\alpha > 1$ and $-1 < \alpha^* < 0$.

(a) If α has the given continued fraction, explain why $\alpha > 1$ (easy). Find the minimal polynomial of α in terms of the numerators and denominators of convergents, and use it to show that $-1 < \alpha^* < 0$.

(b) Now suppose that $\alpha > 1$ and $-1 < \alpha^* < 0$. We know from Theorem 4.9 that the continued fraction

$$\alpha = a_0 + \cfrac{1}{a_1 +} \cfrac{1}{a_2 +} \cdots$$

is *ultimately* periodic: that is, there are integers n and m_0 such that $a_{m+n} = a_m$ whenever $m \geq m_0$. As usual we write α_k for the kth complete quotient of α.

First, show that for all $k \geq 0$, we have $\alpha_k > 1$ and $-1 < \alpha_k^* < 0$.

(c) Next, prove that for every k, we have

$$a_k = \left\lfloor -\frac{1}{\alpha_{k+1}^*} \right\rfloor ,$$

where $\lfloor \cdot \rfloor$ is the floor or greatest–integer function.

(d) Finally, show that $a_{m_0-1+n} = a_{m_0-1}$, and explain how this proves the stated result.

4.9 Let d be a non–square positive integer. Show that if x, y are positive integers and $x^2 - dy^2 = \pm 1$, then x/y is a convergent to $\sqrt{d}$.

4.10 (a) Let $a_1, a_2, \ldots, a_n$ be positive integers, and let p_k/q_k be the kth convergent (in lowest terms) of the continued fraction

$$\cfrac{1}{a_1 +} \cfrac{1}{a_2 +} \cdots \cfrac{1}{a_n} .$$

Show that if $a_n > 1$, then the continued fraction

$$\alpha = \cfrac{1}{a_1 +} \cdots \cfrac{1}{a_{n-1} +} \cfrac{1}{(a_n + 1) +} \cfrac{1}{(a_n - 1) +} \cfrac{1}{a_{n-1} +} \cdots \cfrac{1}{a_1}$$

is equal to

$$\frac{p_n}{q_n} + \frac{(-1)^n}{q_n^2}.$$

(b) Show how one may find, without computing assistance, the partial quotients of the "decimal"

$$\beta = 0.110100010000000100\cdots = \sum_{k=0}^{\infty} \frac{1}{g^{2^k}}$$

in base $g > 2$. Illustrate your solution by finding the 2345th partial quotient of β.

(c) Find the exact order of approximability of β. That is, find s such that β is approximable to order s, and to no higher order.

(d) What happens if $g = 2$?

4.11 Prove that the continued fraction

$$\alpha = \cfrac{1}{10^{1!} +} \cfrac{1}{10^{2!} +} \cfrac{1}{10^{3!} +} \cdots$$

represents a Liouville number.

4.12 (a) Prove *Kronecker's approximation theorem*: for any real irrational α, any real β and any positive real ε there exist infinitely many pairs of integers p, q with $q > 0$ such that $|q\alpha - p - \beta| < \varepsilon$.

(b) Show that the following is equivalent to Kronecker's Theorem: for any real irrational α and any real β_1, β_2 with $\beta_1 < \beta_2$, there exist infinitely many pairs p, q with $q > 0$ such that

$$\beta_1 < q\alpha - p < \beta_2.$$

(c) Use the result in (b) to prove that there is a power of 2 beginning with any given string of decimal digits.

4.13 Use continued fractions to "explain" the approximation

$$\pi \approx \sqrt[4]{\frac{2143}{22}} = 3.14159265258\cdots,$$

discovered by Ramanujan. Also the approximation

$$\pi \approx \left(\frac{16}{9}\right)^2 = 3.16049\cdots:$$

this is implicit in the Rhind papyrus from ancient Egypt, which was copied in about 1500 BC from an earlier source.

4.14 Let α be irrational. Prove that at least one of any two consecutive convergents to α satisfies the inequality

$$\left| \alpha - \frac{p}{q} \right| < \frac{1}{2q^2} \, .$$

4.15 To ten decimal places we have $\pi^{12}/\zeta(12) = 924041.7872648336$. Use continued fractions to conjecture a rational value for $\pi^{12}/\zeta(12)$.

Comment. To get a convincing answer you will need to use a calculator with at least 10 digits accuracy.

4.16 For any $n \geq 1$, find the best possible constant A_n such that the following result is true: if α is a real irrational number with infinitely many partial quotients $a_k \geq n$, then the inequality

$$\left| \alpha - \frac{p}{q} \right| < \frac{1}{A_n q^2}$$

has infinitely many rational solutions p/q.

4.17 Show that the polynomial

$$\begin{aligned} f(z) &= 80(z-1)(10z-11)(9z-10) - 1 \\ &= 7200z^3 - 23120z^2 + 24720z - 8801 \end{aligned}$$

has three real positive roots, and find the first seven partial quotients of the continued fraction of the middle one.

4.18 Let $\alpha = \tan(\pi/5)$. Use the minimal polynomial of α to calculate some of the partial quotients in the continued fraction of α, and hence find a rational number p/q such that

$$\left| \tan\left(\frac{\pi}{5} \right) - \frac{p}{q} \right| < \frac{1}{50q^2} \, .$$

4.19 Show that if $k \geq 2$, then

$$e^{1/k} = 1 + \frac{1}{k-1+} \; \frac{1}{1+} \; \frac{1}{1+} \; \frac{1}{3k-1+} \; \frac{1}{1+} \; \frac{1}{1+} \; \frac{1}{5k-1+} \; \cdots \, .$$

4.20 Let α be an irrational number with partial quotients a_k; suppose that there exists a constant A such that $a_k \leq Ak$ whenever $k \geq 1$. Show that there is a constant c such that

$$\left| \alpha - \frac{p}{q} \right| > \frac{c}{q^2 \log q}$$

for all rational numbers p/q with $q > 1$. Deduce that any such α, and in particular $\alpha = e$, is not approximable to order greater than 2.

4.21 Let ν be a positive real number. The *modified Bessel function of the first kind* of order ν (compare problem 2.6) is defined by the power series

$$I_\nu(x) = \sum_{k=0}^{\infty} \frac{1}{k!\,\Gamma(\nu+k+1)} \left(\frac{x}{2}\right)^{2k+\nu} ;$$

the factor $\Gamma(\nu+k+1)$ in the denominator is a value of the gamma function, which has the property

$$\Gamma(x+1) = x\Gamma(x)$$

for all $x > 0$. Express the continued fraction

$$1 + \frac{1}{2+}\ \frac{1}{3+}\ \frac{1}{4+}\ \frac{1}{5+}\ \cdots$$

in terms of modified Bessel functions.

4.22 *An alternative derivation of the continued fraction for* e. Let α be the real number with continued fraction

$$\alpha = 2 + \frac{1}{1+}\ \frac{1}{2+}\ \frac{1}{1+}\ \frac{1}{1+}\ \frac{1}{4+}\ \frac{1}{1+}\ \cdots ;$$

write a_k for its partial quotients and p_k/q_k for its convergents, and for any k set $r_k = p_k - q_k e$. Define

$$I_k = \int_0^1 \frac{x^{k+1}(x-1)^k}{k!}\, e^x\, dx$$

$$J_k = \int_0^1 \frac{x^k(x-1)^{k+1}}{k!}\, e^x\, dx$$

$$K_k = \int_0^1 \frac{x^{k+1}(x-1)^{k+1}}{(k+1)!}\, e^x\, dx .$$

Show that

$$I_k = r_{3k-1}, \qquad J_k = r_{3k}, \qquad K_k = -r_{3k+1}$$

for $k \geq 0$, and that $I_k, J_k, K_k \to 0$ as $k \to \infty$; deduce that $\alpha = e$.

4.23 *Using continued fractions to break an RSA code.* In RSA encryption, a modulus n and an exponent e are made available publicly. A message m is encoded as $c \equiv m^e \pmod{n}$. The modulus is the product of two large primes, $n = pq$, where p and q are kept secret. Those who know p and q can calculate $\phi(n) = (p-1)(q-1)$ and decode the message by calculating $m \equiv c^d \pmod{n}$: here d is the inverse of e modulo $\phi(n)$, that is, $de \equiv 1 \pmod{\phi(n)}$. It is commonly asserted that the system is safe because finding d is "essentially" equivalent to factorising n, a computationally difficult task. However, care must be taken with the choice of d.

(a) Prove that if $e < (p-1)(q-1)$ and $p < q < 2p$ and $(3d)^4 < n$, then d is the denominator of one of the convergents to e/n.

(b) Suppose that the conditions in (a) hold. Explain why there are only, more or less, $\log n$ candidates for d, and why this makes the encryption insecure.

(c) Implement these ideas in the following small–scale example. Suppose that the public parameters of the code are

$$n = 376146669038857 \quad \text{and} \quad e = 7654913878769$$

(though in a real–life situation, n would have 200 digits or more). If p and q satisfy $p < q < 2p$, and if $d \le 1467$, break the code by determining d.

4.24 Consider an $a \times b$ rectangle, with $a \le b$. By *reducing* such a rectangle we mean cutting off as many as possible $a \times a$ squares, beginning at the side of length a. Suppose we begin with a rectangle of size $1 \times \alpha$, with $\alpha > 1$; reduce it (we call this the 0th step) to obtain a smaller rectangle; reduce this (the 1st step) to obtain a smaller rectangle again; and so on.

(a) Determine how many squares are cut off at the kth step.

(b) Find the size of the rectangle remaining after the kth step.

(c) If $\alpha = 1.2345$, draw an accurate scale diagram of the process up to the fourth step.

4.25 "We are going well," said [Sherlock Holmes], looking out the window and glancing at his watch. "Our rate at present is fifty–three and a half miles an hour." "I have not observed the quarter–mile posts," said [Watson]. "Nor have I. But the telegraph posts upon this line are sixty yards apart, and the calculation is a simple one." (Sir Arthur Conan Doyle, *The Adventure of Silver Blaze*.)

What, exactly, do you think was the calculation performed by Holmes? Give reasons for your opinion.

4.26 An investment company advertises (*Sydney Morning Herald*, 7 September 2002) "the potential to [earn] over 20%". A footnote explains that "20% or more was achieved in 29.41% of simulated tests". Can you find any reason to doubt the integrity of this company?

APPENDIX 1: A PROPERTY OF POSITIVE FRACTIONS

A simple property of positive fractions. If a, b, c and d are positive and a/b is not equal to c/d, then

$$\frac{a+c}{b+d}$$

lies (strictly) between a/b and c/d.

APPENDIX 2: SIMULTANEOUS EQUATIONS WITH INTEGRAL COEFFICIENTS

Let a, b, c, d and p, q be integers. If $ad - bc = \pm 1$, then the simultaneous equations

$$\begin{cases} ax + by = p \\ cx + dy = q \end{cases}$$

have an integral solution x, y. Conversely, if the system has an integral solution for *all* integers p, q, then $ad - bc = \pm 1$.

Proof. The solution can be written

$$\begin{pmatrix} x \\ y \end{pmatrix} = \begin{pmatrix} a & b \\ c & d \end{pmatrix}^{-1} \begin{pmatrix} p \\ q \end{pmatrix} = \frac{1}{ad - bc} \begin{pmatrix} d & -b \\ -c & a \end{pmatrix} \begin{pmatrix} p \\ q \end{pmatrix} = \frac{1}{ad - bc} \begin{pmatrix} dp - bq \\ aq - cp \end{pmatrix}$$

provided that $ad - bc \neq 0$. It is clear that if $ad - bc = \pm 1$, then x and y are integers. Conversely, suppose that $|ad - bc| > 1$ and consider the solutions when $p = 1$, $q = 0$ and when $p = 0$, $q = 1$. If these solutions are to be integers, then $ad - bc$ must be a factor of a, b, c and d. But this leads to

$$(ad - bc)^2 \mid ad - bc \, ,$$

which is impossible. Finally note that if $ad - bc = 0$, then there exist p, q for which the system has no solution at all, and therefore certainly no integral solution.

Exercise. Let A be an $n \times n$ matrix with integral entries. Show that the linear equations $A\mathbf{x} = \mathbf{b}$ have a solution $\mathbf{x}$ with integral components for all integer vectors $\mathbf{b}$, if and only if $\det(A) = \pm 1$.

APPENDIX 3: CARDINALITY OF SETS OF SEQUENCES

Theorem 4.23. *Let A be a set with more than one element. Then the set*

$$S = \{ (a_0, a_1, a_2, \ldots) \mid a_k \in A \text{ for all } k \}$$

of sequences in A is uncountable.

Proof. Suppose that S is countable; then by definition (see Chapter 3, appendix 1) there is a one–to–one function f from S to $\mathbb{N}$. Let x and y be distinct elements of A. Since f is one–to–one each $k \in \mathbb{N}$ is the image of at most one sequence in S, and so there is a well–defined sequence $(b_0, b_1, b_2, \ldots)$ given by

$$b_k = \begin{cases} x & \text{if } k = f(a_0, a_1, a_2, \ldots) \text{ and } a_k = y \\ y & \text{otherwise;} \end{cases}$$

note that the "otherwise" includes both the case where a_k is an element of A other than y and the case where k is not in the range of f. Clearly $(b_0, b_1, b_2, \ldots)$ is in S, and so $f(b_0, b_1, b_2, \ldots)$ is equal to some natural number k. But this is impossible since from the definition we have $b_k = x$ if $b_k = y$, and $b_k = y$ if $b_k \neq y$. We have a contradiction, and the result is proved.

APPENDIX 4: BASIC MUSICAL TERMINOLOGY

Some musical terminology, for readers who may not already be familiar with it. Consider a piano keyboard, part of which is shown in figure 4.3. The white

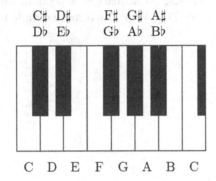

Figure 4.3 Part of a piano keyboard.

keys are labelled with the first seven letters of the alphabet: A, B, C, D, E, F, G. After using all of these we start again. Note that A follows G, and that C appears at both the left and right–hand ends of the diagram. A black key is given the name of the white key just below it, with a *sharp* (♯) added; or of the white key just above, with a *flat* (♭) added. Thus the leftmost black key in the diagram is called C♯ or D♭, pronounced "C sharp", "D flat".

The interval from any key to the next is called a *semitone*. This is the smallest interval used in the majority of traditional Western music. For example, C–C♯, G♯–A, B–C are all semitones. The interval from any key to the next key of the same name (for example C–C, E♭–E♭) is called an *octave*. Counting five steps up a scale (including the first and last notes) gives the interval of a *fifth*, sometimes, for emphasis, called a *perfect fifth*. (Musical readers will know that there are other kinds of fifths, but we shall not be concerned with them here.) Instances of perfect fifths are C–G, B–F♯ and G♭–D♭.

Musical sounds are caused by regular vibrations in the air (or in other media). The number of vibrations per second causing any particular note is the *frequency* of that note in units of *Hertz* (Hz). For example, the modern standard of orchestral pitch is established by *defining* the note A above middle C to have a frequency of exactly 440 Hz – that is, 440 vibrations per second. It is found by observation (and backed up by psychological and physiological theories) that when two notes of different pitches are sounded simultaneously or consecutively, the result is most pleasing to the ear if the *ratio* of the frequencies of the pitches is a simple fraction. The simplest possible fractions are $\frac{2}{1}$ and $\frac{3}{2}$, and these correspond to the intervals of the octave and the perfect fifth respectively. Middle C, for example, has a frequency of 262 Hz; a perfect fifth above is G with a frequency of 393 Hz; the octave above middle C is the C with frequency 524 Hz.

The above is adapted, with permission, from the present author's article [3]. For accessible reading on acoustical aspects of music, the classic text is that by Sir James Jeans [34]. Very much more detailed information may be found in [12].

CHAPTER 5

Hermite's Method for Transcendence

Be it enacted by the General Assembly of the State of Indiana:
It has been found that a circular area is to the square on
a line equal to the quadrant of the circumference, as the
area of an equilateral rectangle is to the square on one side...
The present rule... is entirely wrong...

House Bill No. 246 (1897), State of Indiana[1]

Now I, even I, would celebrate
In rhymes unapt, the great
Immortal Syracusan, rivaled nevermore,
Who in his wondrous lore
Passed on before
Left men his guidance
How to circles mensurate.

Adam C. Orr [48]

A S POINTED OUT IN CHAPTER 2, it is often easy to prove the irrationality of a number specifically constructed so as to be irrational, but can be much harder to prove a given "naturally occurring" number irrational. The same remarks apply with still more force to the question of transcendence: we have already shown that the number

$$\alpha = \sum_{k=1}^{\infty} \frac{1}{10^{k!}} = 0.110001000000000000000001000\cdots$$

---

[1]In 1897 the parliament of the US state of Indiana was presented with a bill which would, in effect, have fixed by legislation the value of π. If taken literally, the section quoted is equivalent to the formula $A = (\frac{1}{4}c)^2$ for the area of a circle, which implies that $\pi = 4$. The bill was passed unanimously by the Lower House and sent to the Senate, where owing to the fortunate intervention of a professor of mathematics, its further consideration was postponed indefinitely. See Petr Beckmann, *A History of π* [11], Chapter 17.

DOI: 10.1201/9781003111207-5

is transcendental, but this is hardly a number which one would expect to encounter in any other area of mathematics. In the present chapter we shall demonstrate the transcendence of two of the most important constants of mathematics, e and π, and shall use the same techniques, but in a more complex way, to prove an important theorem of Lindemann which generalises both of these results. We shall also develop some properties of symmetric polynomials, which will be required in proving the transcendence of π.

5.1 TRANSCENDENCE OF e

The proof of the transcendence of e is in fact not very different from Hermite's proof of the irrationality of e^r, though naturally the details are more complicated. As in Chapter 2 we'll try to provide some motivation before giving a formal argument. One might expect the proof to be by contradiction, and so we begin by assuming that e is algebraic: thus, there is a polynomial identity

$$a_m e^m + a_{m-1} e^{m-1} + \cdots + a_1 e + a_0 = 0 , \tag{5.1}$$

where the coefficients a_k are integers. Our arguments in Chapter 2 were based upon the integral formula

$$\int_0^r f(x) e^x \, dx = F(r) e^r - F(0) e^0$$

for a certain function F. If we try to employ the same expression in a transcendence proof, we find that the product $F(r)e^r$ of two "variable" or "unknown" quantities causes difficulties. In Chapter 2, the assumption $e^r = p/q$ simplified the formula to such an extent that we were able to complete the proof; for future work, however, it will be advantageous to have an expression of the form $F(r)e^0 - F(0)e^r$ in which we can deal with the two difficulties separately. To interchange the roles of 0 and r in the exponential we need only integrate a function of the form $f(x)e^{r-x}$ instead of $f(x)e^x$.

We should also give some thought to the range of integration. Integrating from 0 to r was successful in Chapter 2 since we had information concerning e^0 (of course!) and e^r (by assumption). In the present case our assumption (5.1) involves $e^0, e^1, e^2, \ldots, e^m$, and so we shall consider the $m + 1$ integrals

$$I_k = \int_0^k f(x) e^{k-x} \, dx ,$$

where $k = 0, 1, 2, \ldots, m$ and f is a polynomial to be chosen later.

Comment. Assuming integrability over a suitable interval, define the *convolution* of functions f and g to be the function given by

$$(f * g)(x) = \int_0^x f(t) g(x - t) \, dt .$$

Then (exercise!) the operation $*$ is associative, $(f * g) * h = f * (g * h)$, while the operation defined by

$$(f \odot g)(x) = \int_0^x f(t)g(t)\,dt$$

is not; this fact alone suggests that, despite appearances, $*$ is a more "mathematically natural" way of combining multiplication and integration than is $\odot$. If we denote the exponential function by exp, then the integrals we are now considering are certain values of $f * \exp$, while those we used in Chapter 2 were related to the less important $f \odot \exp$.

The convolution of two functions appears in the study of Laplace transforms; a slightly different type of convolution has connections with Fourier transforms. The **Dirichlet product**

$$(f * g)(n) = \sum_{d|n} f(d)g\left(\frac{n}{d}\right)$$

of two arithmetic functions f and g is very important in number theory, and may be seen as a discrete analogue of the convolution.

We return to the development of a transcendence proof for e. Integrating by parts, we obtain

$$I_k = F(0)e^k - F(k)\ ,$$

where

$$F(x) = f(x) + f'(x) + f''(x) + \cdots\ .$$

Our transcendence proof will rely on the expression

$$J = \sum_{k=0}^m a_k I_k = -\sum_{k=0}^m a_k F(k)\ .$$

We shall aim to show by estimating the integrals I_k that J is "small", and by analysing the derivatives of f that J is a non–zero integer and therefore "large"; this will give the sort of contradiction that we have seen a number of times in Chapter 2, and will prove that e is not algebraic.

To make $F(k)$ simple for $k = 0, 1, 2, \ldots, m$ we shall choose $f(x)$ to have many factors of $x - k$ for each k. Recall that (perhaps surprisingly) one of the more intricate aspects of the proofs in Chapter 2 was the necessity of showing an integral expression such as J to be non–zero. Our earliest proofs relied on the fact that the integrand was positive; but here we know essentially nothing about the coefficients a_k, and so this method appears unlikely to succeed. We take inspiration, instead, from the argument employed in the proof of Theorem 2.5 (page 25), where we proved that a certain expression was non–zero because it was not a multiple of $(n + 1)!$. This worked because f was divisible by a high power of x, and an *even higher* power of $x - r$. So we shall set

$$f(x) = x^n (x - 1)^{n+1} (x - 2)^{n+1} \cdots (x - m)^{n+1}$$

where, as usual, n is to be chosen later.

The above ideas will suffice to prove e transcendental. We shall need the lemma on derivatives of polynomials from Chapter 2; for convenience of reference we restate (a particular case of) this lemma.

Lemma 5.1. Derivatives of polynomials. *Let a be an integer, n a non-negative integer, and g a polynomial with integral coefficients. Define the polynomial f by*

$$f(x) = (x - a)^n g(x) \ .$$

Then for all $j \geq 0$, the derivative $f^{(j)}(a)$ is an integer divisible by $n!$.

Theorem 5.2. (Hermite, 1873). *The exponential constant e is transcendental.*

Proof. Suppose that e is an algebraic number of degree m, and therefore satisfies an algebraic equation

$$a_m e^m + a_{m-1} e^{m-1} + \cdots + a_1 e + a_0 = 0 \ . \tag{5.2}$$

Without loss of generality we may assume that the coefficients a_k are rational integers with $a_0 \neq 0$. For any positive integer n let

$$f(x) = x^n (x - 1)^{n+1} (x - 2)^{n+1} \cdots (x - m)^{n+1} \ ,$$

and define

$$I_k = \int_0^k f(x) e^{k-x} \, dx$$

for $k = 0, 1, 2, \ldots, m$. Integrating repeatedly by parts (exercise!) we obtain

$$I_k = F(0) \, e^k - F(k) \ , \tag{5.3}$$

where

$$F(x) = f(x) + f'(x) + f''(x) + \cdots \ .$$

For $k = 1, 2, \ldots, m$ the lemma shows that $(n + 1)! \mid f^{(j)}(k)$ for all $j \geq 0$, and hence that $(n + 1)!$ is a factor of $F(k)$. To handle the case $k = 0$ we note that

$$f(x) = (-1)^{m(n+1)} (m!)^{n+1} x^n + \{\text{ higher order terms }\}$$

and so

$$f^{(j)}(0) = j! \times \{\text{ coefficient of } x^j \text{ in } f(x) \}$$

$$= \begin{cases} 0 & \text{if } j < n \\ (-1)^{m(n+1)} (m!)^{n+1} n! & \text{if } j = n \\ \text{a multiple of } (n + 1)! & \text{if } j > n. \end{cases}$$

Now set

$$J = a_0 I_0 + a_1 I_1 + a_2 I_2 + \cdots + a_m I_m .$$

Using equations (5.2) and (5.3), we have

$$J = \sum_{k=0}^{m} a_k \left(F(0) e^k - F(k) \right) = - \sum_{k=0}^{m} a_k F(k) ,$$

and the divisibility properties we have just proved show that

$$J = (-1)^{mn+m+1} (m!)^{n+1} n! \, a_0 + \{ \text{a multiple of } (n+1)! \} . \tag{5.4}$$

On the other hand, we can find an upper bound for J by using the usual sort of integral estimate. First, observe that the range of integration for every I_k is a subset of the interval $0 \le x \le m$. Thus for all relevant x the polynomial $f(x)$ is a product of $mn + m + n$ factors, each of absolute value at most m, so

$$|I_k| \le \int_0^k |f(x)| e^{k-x} \, dx \le k m^{mn+m+n} e^k \le m^{mn+m+n+1} e^m$$

and hence

$$|J| \le \sum_{k=0}^{m} |a_k I_k| \le \left(\sum_{k=0}^{m} |a_k| \right) m^{m+1} e^m \left(m^{m+1} \right)^n .$$

Since we have assumed e to be algebraic, its degree m and the coefficients of its minimal polynomial are fixed numbers, independent of n, and this inequality can be written

$$|J| \le ab^n \tag{5.5}$$

with a and b independent of n. To complete the proof, choose n such that $n+1$ is a prime number greater than both m and $|a_0|$, and large enough that $n! > ab^n$. Then (5.4) shows that J is an integer which is divisible by $n!$ but not by $(n+1)!$; thus $|J| \ge n! > ab^n$, which contradicts (5.5). Therefore, the assumption that e is algebraic is untenable, and we have shown that e is transcendental.

Corollary 5.3. *If r is a non–zero rational number, then e^r is transcendental.*

Proof. Let $\beta = e^r = e^{p/q}$. If β is algebraic, then e is a root of the polynomial equation

$$z^p - \beta^q = 0$$

with algebraic coefficients; hence, by Theorem 3.12, the number e is algebraic. But we have just shown that this is not so.

Comment. Another interesting question is whether or not e^α need be transcendental for an *algebraic* number $\alpha \ne 0$. We shall return to this later.

5.2 TRANSCENDENCE OF π

We now turn to proving the transcendence of π. Some features of the proof will be very similar to the one we have just done: the underlying reason for this is that π is closely connected with the exponential function by virtue of Euler's formula $e^{i\pi} = -1$. To take advantage of this connection, however, we shall have to consider not only real but also complex algebraic numbers. The main additional difficulty we shall encounter will be in constructing the polynomial f and hence the integrals I_k. We shall assume that π is algebraic, and use this assumption to construct a polynomial with known roots; but we shall need to prove that this polynomial has rational coefficients. To do so we shall need certain facts about *symmetric polynomials*; these facts are connected with the well–known relations between roots and coefficients of a polynomial.

5.2.1 Symmetric polynomials

Definition 5.1. *A* **symmetric polynomial** *f in m variables is a polynomial with the property that if $(y_1, y_2, \ldots, y_m)$ is a permutation of $(x_1, x_2, \ldots, x_m)$, then*

$$f(y_1, y_2, \ldots, y_m) = f(x_1, x_2, \ldots, x_m) \ .$$

The **elementary symmetric polynomials** *in m variables are the polynomials e_k for $k = 0, 1, 2, \ldots, m$, where $e_k(x_1, x_2, \ldots, x_m)$ is the sum of all products of k distinct variables from $\{x_1, x_2, \ldots, x_m\}$.*

Examples. The elementary symmetric polynomials in x_1, x_2, x_3, x_4 are

$$e_0 = 1 \ , \quad e_1 = x_1 + x_2 + x_3 + x_4 \ ,$$
$$e_2 = x_1x_2 + x_1x_3 + x_1x_4 + x_2x_3 + x_2x_4 + x_3x_4 \ ,$$
$$e_3 = x_1x_2x_3 + x_1x_2x_4 + x_1x_3x_4 + x_2x_3x_4 \ , \quad e_4 = x_1x_2x_3x_4 \ .$$

Polynomials such as

$$f(x_1, x_2, x_3, x_4) = x_1^2 + x_2^2 + x_3^2 + x_4^2 - 7x_1 - 7x_2 - 7x_3 - 7x_4$$

and

$$f(x_1, x_2, x_3, x_4) = \left\{ \begin{array}{l} x_1^2x_2^2x_3 + x_1^2x_2^2x_4 + x_1^2x_2x_3^2 + x_1^2x_3^2x_4 \\ + x_1^2x_2x_4^2 + x_1^2x_3x_4^2 + x_1x_2^2x_3^2 + x_2^2x_3^2x_4 \\ + x_1x_2^2x_4^2 + x_2^2x_3x_4^2 + x_1x_3^2x_4^2 + x_2x_3^2x_4^2 \end{array} \right\} \quad (5.6)$$

are not *elementary* symmetric polynomials, but they are symmetric because any reordering of the variables will return the same expression. The polynomial

$$f(x_1, x_2, x_3, x_4) = x_1^2x_2^2x_3 + x_2^2x_3^2x_4 + x_3^2x_4^2x_1 + x_4^2x_1^2x_2$$

is not symmetric since $f(x_1, x_2, x_3, x_4) \neq f(x_2, x_1, x_3, x_4)$.

The elementary symmetric polynomials, as suggested above, form the basis of the well–known relations between roots and coefficients of a polynomial.

Lemma 5.4. *Let $e_0, e_1, e_2, \ldots, e_m$ be the elementary symmetric polynomials in $\alpha_1, \alpha_2, \ldots, \alpha_m$. Then*

$$(1 + \alpha_1)(1 + \alpha_2) \cdots (1 + \alpha_m) = e_0 + e_1 + e_2 + \cdots + e_m ,$$

and for any x, we have

$$(x - \alpha_1)(x - \alpha_2) \cdots (x - \alpha_m) = x^m - e_1 x^{m-1} + e_2 x^{m-2} - \cdots + (-1)^m e_m .$$

Proof. For the first identity, just stare at it until it becomes obvious! Then replace α_k by $-\alpha_k/x$, observe that

$$e_k\left(-\frac{\alpha_1}{x}, -\frac{\alpha_2}{x}, \ldots, -\frac{\alpha_m}{x}\right) = (-1)^k x^{-k} e_k(\alpha_1, \alpha_2, \ldots, \alpha_m) ,$$

and multiply through by x^m; this proves the second result.

One reason for the importance of the elementary symmetric polynomials is that they can be used to write an expression for *any* symmetric polynomial.

Theorem 5.5. Symmetric polynomials. *Let f be a symmetric polynomial in $x_1, x_2, \ldots, x_m$, with coefficients in R, an additive subgroup of the real numbers. Let e_k be the elementary symmetric polynomials in $x_1, x_2, \ldots, x_m$. Then there is a polynomial g in m variables, having coefficients in R and degree at most that of f, such that*

$$f(x_1, x_2, \ldots, x_m) = g(e_1, e_2, \ldots, e_m) .$$

Comment. More tersely stated: any symmetric polynomial is a polynomial of the same or smaller degree in the elementary symmetric polynomials; and if the former polynomial has coefficients in R, then so does the latter.

Proof by induction on n, the degree of f, and m, the number of variables. The result is trivial if $n = 0$ or $m = 1$; let f be a symmetric polynomial of degree n in $x_1, \ldots, x_m$, having coefficients in R, and assume that the result is true for all polynomials with degree at most n in at most m variables (with at least one of these inequalities strict). Let

$$f^*(x_1, \ldots, x_{m-1}) = f(x_1, \ldots, x_{m-1}, 0) ,$$

and similarly write

$$e_k^* = e_k(x_1, \ldots, x_{m-1}, 0) ;$$

it is not hard to see that the e_k^* are in fact the elementary symmetric polynomials in $x_1, \ldots, x_{m-1}$.

Now since f is symmetric, f^* is also symmetric; it has fewer than m variables and degree at most n, and so by the inductive assumption there is a polynomial g^* with coefficients in R, such that $f^* = g^*(e_1^*, \ldots, e_{m-1}^*)$. Consider the polynomial

$$f(x_1, \ldots, x_m) - g^*(e_1, \ldots, e_{m-1}) \ ,$$

which is clearly symmetric in $x_1, \ldots, x_m$. This polynomial has x_m as a factor; since it is symmetric, each other x_k is also a factor, and hence e_m is a factor. Thus

$$f(x_1, \ldots, x_m) - g^*(e_1, \ldots, e_{m-1}) = e_m h(x_1, \ldots, x_m) \ ,$$

where h is again a symmetric polynomial. Since e_k and e_k^* are of equal degree for $k = 1, 2, \ldots, m - 1$, we have

$$\deg\big(g^*(e_1, \ldots, e_{m-1})\big) = \deg\big(g^*(e_1^*, \ldots, e_{m-1}^*)\big) = \deg f^* \le \deg f \ ,$$

where in the first two terms the degrees are in terms of the variables $x_1, \ldots, x_m$, and so h has smaller degree than f; also, h has coefficients in R. By induction we may assume that h is a polynomial in $e_1, \ldots, e_m$ with coefficients in R, and since g^* also has coefficients in R, so does

$$g(e_1, \ldots, e_m) = g^*(e_1, \ldots, e_{m-1}) + e_m h(x_1, \ldots, x_m) \ .$$

Finally, every e_k has degree at least 1 in the variables $x_1, \ldots, x_m$; so g cannot have degree exceeding that of f, and this completes the proof.

Comments.

- In applying the above theorem we shall mainly be interested in three cases: $R = \mathbb{Q}$, the set of rational numbers; $R = \mathbb{Z}$, the set of integers; and $R = c\mathbb{Z}$, the set of multiples of a fixed integer c.

- The proof provides an algorithm for expressing a symmetric polynomial in terms of elementary symmetric polynomials. For example, take f to be the polynomial in four variables given by (5.6) on page 114. Then

$$f^* = x_1^2 x_2^2 x_3 + x_1^2 x_2 x_3^2 + x_1 x_2^2 x_3^2 \ ,$$

and it is easy to see that $f^* = e_2^* e_3^*$. (In a more difficult case we would iterate the algorithm to give $f^{**} = 0$, and so by the theorem $f^* = e_3^* h^*$ with $h^* = x_1 x_2 + x_1 x_3 + x_2 x_3$; if it were not clear how to express h^* in terms of elementary symmetric functions we would now apply the algorithm to h^*.) Therefore, the theorem tells us that e_4 is a factor of $f - e_2 e_3$; performing the algebra, we find

$$\begin{aligned} f - e_2 e_3 &= -3x_1^2 x_2 x_3 x_4 - 3x_1 x_2^2 x_3 x_4 - 3x_1 x_2 x_3^2 x_4 - 3x_1 x_2 x_3 x_4^2 \\ &= -3e_4(x_1 + x_2 + x_3 + x_4) \\ &= -3e_1 e_4 \ , \end{aligned}$$

and so

$$f = e_2 e_3 - 3e_1 e_4 \ ,$$

which, as claimed, is a polynomial in the four variables e_1, e_2, e_3, e_4.

- For a different algorithm (and, consequently, a different proof of the preceding theorem) see Stewart and Tall [62], pages 24–27.

Corollary 5.6. Evaluations of symmetric polynomials.

- *Let f be a monic polynomial of degree m with integer coefficients. Suppose that f has roots $\alpha_1, \alpha_2, \ldots, \alpha_m$ (including repeated roots, if any) and let c be a symmetric polynomial in m variables with coefficients in R, an additive subgroup of the real numbers. Then $c(\alpha_1, \alpha_2, \ldots, \alpha_m)$ is an element of R.*

- *If R is $\mathbb{Q}$, or indeed any subfield of $\mathbb{R}$, then the above holds for all polynomials f, monic or not.*

Proof. To prove the first statement we note that from the preceding theorem, $c(\alpha_1, \alpha_2, \ldots, \alpha_m)$ is a polynomial in the elementary symmetric polynomials of $\alpha_1, \alpha_2, \ldots, \alpha_m$, having coefficients in R; but these elementary symmetric polynomials are (up to sign) the coefficients of f, and hence are integers. Finally, a polynomial with coefficients in R, evaluated at integer arguments, is a sum of integer multiples of elements of R, and hence belongs to the same subgroup.

The argument for the second statement is almost identical, only noting that in this case the elementary symmetric polynomials of $\alpha_1, \alpha_2, \ldots, \alpha_m$ are the coefficients of f, divided by its leading coefficient, and hence are rational; and any rational multiple of an element of the field R is also in R.

5.2.2 The transcendence proof

The only further preparation we need before embarking on the transcendence proof for π is to recall that the *conjugates* of an algebraic number α are the roots of its minimal polynomial; if α is of degree m, then it has m conjugates, one of which is α itself.

Theorem 5.7. (Lindemann, 1882): *π is transcendental.*

Proof. Suppose that π is algebraic; then $i\pi$, being a product of algebraic numbers, is also algebraic. Let m be the degree of $i\pi$, and let the conjugates of $i\pi$ be $\alpha_1, \alpha_2, \ldots, \alpha_m$. We have

$$(e^{\alpha_1} + 1)(e^{\alpha_2} + 1) \cdots (e^{\alpha_m} + 1) = 0 \tag{5.7}$$

because one of the factors is $e^{i\pi} + 1$. Expanding the left–hand side we obtain a sum of 2^m terms e^{β_S}, where for any $S \subseteq \{1, 2, \ldots, m\}$ we set

$$\beta_S = \sum_{k \in S} \alpha_k .$$

That is, the values of β_S include all α_k, all sums $\alpha_{k_1} + \alpha_{k_2}$ with $k_1 \neq k_2$, and in general all sums of any number of distinct α_k. Note that this includes the empty sum $\beta_\varnothing = 0$, which corresponds to the product $1 \times 1 \times \cdots \times 1$; it is possible that other sums β_S are also zero.

Ferdinand von Lindemann
(1852–1939)

Let g be the monic polynomial of degree 2^m whose roots are the numbers β_S, including any multiplicities. That is,

$$g(z) = \prod_{S \subseteq \{1,2,\ldots,m\}} (z - \beta_S) \, .$$

By Lemma 5.4, the coefficients of g are elementary symmetric polynomials in the sums β_S, and therefore can be written as polynomials in the numbers α_k. If the α_k are permuted in any way whatsoever, the expansion of the left–hand side of (5.7) will contain the same terms; therefore the β_S will remain the same, though possibly in a different order; and so the coefficients of g will be unchanged. That is, these coefficients can be written as symmetric polynomials in $\alpha_1, \alpha_2, \ldots, \alpha_m$. But because all of the α_k are conjugates, they are the roots of a single polynomial with rational coefficients, and so, by Corollary 5.6, the polynomial g also has rational coefficients.

Suppose that $g(z)$ has a factor of z with multiplicity s; since $\beta_\varnothing = 0$ is a root of g, we know that $s \geq 1$. Divide g by z^s and multiply by a common denominator for its coefficients to obtain a polynomial

$$h(z) = h_t z^t + h_{t-1} z^{t-1} + \cdots + h_1 z + h_0$$

of degree $t = 2^m - s$, with integral coefficients, of which h_t and h_0 are non–zero. We relabel the non–zero values of β_S, including any repetitions, as $\beta_1, \beta_2, \ldots, \beta_t$; these numbers are therefore the roots of h. Expanding the left–hand side of (5.7), we have

$$e^{\beta_1} + e^{\beta_2} + \cdots + e^{\beta_t} + s = 0 \, . \tag{5.8}$$

From now on the proof follows closely the transcendence proof for e. For any positive integer n, write

$$f(z) = z^n h(z)^{n+1} \, ,$$

a polynomial with integral coefficients and degree less than $(n+1)(t+1)$, and let

$$I_\beta = \int_0^\beta f(z) \, e^{\beta - z} \, dz$$

for any $\beta \in \mathbb{C}$. Note that the integrand of I_β is an entire function. Therefore, we need not specify the path of integration and may take it to be the straight line from 0 to β; moreover, we can compute the integral by using an antiderivative of the integrand. As we have seen many times already,

$$I_\beta = F(0)e^\beta - F(\beta) , \tag{5.9}$$

where $F(z) = f(z) + f'(z) + f''(z) + \cdots$. Now consider

$$J = I_{\beta_1} + I_{\beta_2} + \cdots + I_{\beta_t} = \sum_{k=1}^{t} I_{\beta_k} .$$

Using the results (5.9) and (5.8), and the definition of F, we have

$$J = F(0) \sum_{k=1}^{t} e^{\beta_k} - \sum_{k=1}^{t} F(\beta_k) = -sF(0) - \sum_{j=0}^{\infty} \sum_{k=1}^{t} f^{(j)}(\beta_k) . \tag{5.10}$$

(Remember that the sum over j really has only finitely many terms.) Consider the innermost sum on the right–hand side. Since $f(z)$ contains a factor $z - \beta_k$ with multiplicity $n + 1$ (or possibly more, as the β_k need not all be different), we have

$$f^{(j)}(\beta_k) = 0$$

for $j < n + 1$. Next take $j \geq n + 1$; in this case $f^{(j)}(z)$ is a polynomial with integer coefficients, all divisible by $j!$ and a *fortiori* by $(n + 1)!$. Therefore

$$\sum_{k=1}^{t} f^{(j)}(\beta_k)$$

is a symmetric polynomial in $\beta_1, \beta_2, \ldots, \beta_t$, the coefficients of the polynomial being divisible by $(n+1)!$; so, using the theorem on symmetric polynomials in the case $R = (n+1)!\,\mathbb{Z}$, it can be written as a polynomial with coefficients divisible by $(n+1)!$ in the elementary symmetric polynomials $e_k(\beta_1, \beta_2, \ldots, \beta_t)$. But these elementary symmetric polynomials are the rational numbers $\pm h_{t-k}/h_t$: that is, we can write

$$\sum_{k=1}^{t} f^{(j)}(\beta_k) = p(e_1, e_2, \ldots, e_t) = p\left(-\frac{h_{t-1}}{h_t}, \frac{h_{t-2}}{h_t}, \ldots, \pm\frac{h_0}{h_t}\right) .$$

Since

$$\deg p \leq \deg f^{(j)} \leq (\deg f) - j < (n + 1)t ,$$

multiplying by $h_t^{(n+1)t}$ will clear all the denominators from the right–hand side. Therefore

$$h_t^{(n+1)t} \sum_{k=1}^{t} f^{(j)}(\beta_k)$$

is an integer divisible by $(n+1)!$, and consequently so is

$$h_t^{(n+1)t} \sum_{j=0}^{\infty} \sum_{k=1}^{t} f^{(j)}(\beta_k) = h_t^{(n+1)t} \sum_{k=1}^{t} F(\beta_k) .$$

The evaluation of $F(0)$ is comparatively straightforward: by standard arguments, we have

$$f^{(j)}(0) = \begin{cases} 0 & \text{if } j < n \\ h_0^{n+1} n! & \text{if } j = n \\ \text{a multiple of } (n+1)! & \text{if } j > n. \end{cases}$$

Combining (5.10) with all the divisibility results we have just proved,

$$h_t^{(n+1)t} J = -sh_t^{(n+1)t} h_0^{n+1} n! + \{\text{a multiple of } (n+1)!\} . \tag{5.11}$$

It remains to estimate the integrals I_{β_k}. Let H be the maximum absolute value of the coefficients of h (compare page 41). If z is a complex number lying on the line segment from 0 to β_k, then

$$|h(z)| \leq \sum_{j=0}^{t} |h_j| \, |z|^j \leq H \sum_{j=0}^{t} |\beta_k|^j \leq H(1 + |\beta_k|)^t .$$

Moreover, for the same values of z, we have

$$|e^{\beta_k - z}| = e^{\operatorname{Re}(\beta_k - z)} \leq e^{|\operatorname{Re} \beta_k|} ;$$

putting all this information together,

$$|I_{\beta_k}| \leq |\beta_k| \, |\beta_k|^n H^{n+1} (1 + |\beta_k|)^{t(n+1)} e^{|\operatorname{Re} \beta_k|} .$$

Now $\beta_1, \beta_2, \ldots, \beta_t, H$ and t are all independent of n, so we can write

$$|I_{\beta_k}| \leq ab^n ,$$

where

$$a = a(k) = |\beta_k|(1 + |\beta_k|)^t H e^{|\operatorname{Re} \beta_k|} \quad \text{and} \quad b = b(k) = |\beta_k|(1 + |\beta_k|)^t H$$

depend on k but not on n. One last estimate: if we let A be the greatest of the $a(k)$ for $1 \leq k \leq t$ and B the greatest of the $b(k)$, then

$$\left| h_t^{(n+1)t} J \right| \leq \left| h_t^{(n+1)t} \right| \sum_{k=1}^{t} |I_{\beta_k}| \leq \left| h_t^{(n+1)t} \right| tAB^n = cd^n \tag{5.12}$$

with c and d independent of n, and we have set up the customary contradiction. Choose n such that $n + 1$ is greater than s, greater than the absolute values of both h_0 and h_t, prime, and sufficiently large that $cd^n < n!$. Then from (5.11) we find that $h_t^{(n+1)t}J$ is a non–zero integer divisible by $n!$; using this observation and inequality (5.12), we have

$$n! \leq \left| h_t^{(n+1)t} J \right| \leq cd^n < n! \,,$$

which is a contradiction. Therefore, we have shown that π is transcendental.

Corollary 5.8. *The problem of squaring the circle is unsolvable. That is, it is impossible using ruler and compasses to construct two line segments with lengths in the ratio π.*

Proof. As mentioned in exercise 3.23, it can be proved that segments with lengths in the ratio α can be constructed only if α is an algebraic number whose degree is a power of 2. However, π is not such a number.

5.3 SOME MORE IRRATIONALITY PROOFS

Viewing the transcendence proof for π from a slightly different angle, we have assumed that $\alpha = i\pi$ is algebraic, and have obtained a contradiction by showing that e^α cannot equal the rational number -1. We can use similar ideas to prove that if α is algebraic and non–zero, then e^α cannot equal *any* rational number, or indeed any algebraic number: that is, e^α is transcendental. We shall approach this difficult theorem slowly, beginning by taking a simple specific example, $\alpha = \sqrt{2}$, and seeking only to prove the irrationality of e^α. In the course of the proof we shall show that our careful estimates for $h(z)$ and $e^{\beta_k - z}$ were not really necessary, and may be replaced by an argument based on simple properties of real or complex functions.

Theorem 5.9. $e^{\sqrt{2}}$ *is irrational.*

Comment. In fact, we have already asked the reader to prove this result – see exercise 1.22. However, the method used there, while it also suffices to prove the irrationality of $e^{\sqrt{3}}$, does not appear to generalise any further. The method we now introduce is much more powerful.

Proof. Suppose, on the contrary, that $e^{\sqrt{2}} = p/q$ is rational. Inspired by (5.7), we consider not only $\sqrt{2}$ but also its conjugate $-\sqrt{2}$, and begin by noting that

$$\left(qe^{\sqrt{2}} - p \right)\left(qe^{-\sqrt{2}} - p \right) = 0 \,,$$

because the first factor is zero. Expanding and collecting terms,

$$(p^2 + q^2) - pq\left(e^{\sqrt{2}} + e^{-\sqrt{2}} \right) = 0 \,.$$

Though we shall not use it in the present proof, we note that this equation can be rewritten

$$e^{\sqrt{2}} + e^{-\sqrt{2}} - \frac{p^2 + q^2}{pq} = 0 \,,$$

which is strongly analogous to (5.8). As in the earlier proof, we make use of a polynomial with integral coefficients, having roots $\sqrt{2}$ and $-\sqrt{2}$: clearly $h(z) = z^2 - 2$ is such a polynomial. So we set

$$f(z) = z^n(z^2 - 2)^{n+1}$$

and then, closely following our previous proof,

$$I_\beta = \int_0^\beta f(z)e^{\beta-z}\,dz = F(0)\,e^\beta - F(\beta)$$

with $F = f + f' + f'' + \cdots$, and

$$J = I_{\sqrt{2}} + I_{-\sqrt{2}}\ .$$

Then we have

$$pqJ = pqF(0)\left(e^{\sqrt{2}} + e^{-\sqrt{2}}\right) - pq\left(F(\sqrt{2}) + F(-\sqrt{2})\right)$$
$$= (p^2 + q^2)F(0) - pq\left(F(\sqrt{2}) + F(-\sqrt{2})\right)\ .$$

Now $f(z)$ has factors $z \pm \sqrt{2}$ with multiplicity $n + 1$, and so

$$f^{(j)}(\sqrt{2}) = f^{(j)}(-\sqrt{2}) = 0$$

for any $j < n + 1$. If $j \geq n + 1$, then $f^{(j)}(x_1) + f^{(j)}(x_2)$ is a symmetric polynomial in x_1 and x_2, having coefficients divisible by $(n+1)!$; so it can be written as a polynomial with similar coefficients, evaluated at the elementary symmetric polynomials,

$$f^{(j)}(x_1) + f^{(j)}(x_2) = P(e_1, e_2)\ .$$

If we take $x_1 = \sqrt{2}$ and $x_2 = -\sqrt{2}$, then e_1 and e_2 can be found in terms of the coefficients of $h(z)$ and we have

$$f^{(j)}(\sqrt{2}) + f^{(j)}(-\sqrt{2}) = P(0, -2)\ :$$

this is a multiple of $(n + 1)!$, and hence so is

$$F(\sqrt{2}) + F(-\sqrt{2}) = \sum_{j=0}^\infty \left(f^{(j)}(\sqrt{2}) + f^{(j)}(-\sqrt{2})\right)\ .$$

Moreover,

$$f^{(j)}(0) = \begin{cases} 0 & \text{if } j < n \\ (-2)^{n+1}n! & \text{if } j = n \\ \text{a multiple of } (n + 1)! & \text{if } j > n; \end{cases}$$

we can use these results to evaluate $F(0)$, and hence to obtain

$$pqJ = (-2)^{n+1}(p^2 + q^2)\,n! + \{\,\text{a multiple of } (n + 1)!\,\}\ . \tag{5.13}$$

To estimate the integrals I_β, observe that e^z and $h(z)$ are continuous on the interval $[-\sqrt{2}, \sqrt{2}]$ and hence are bounded there, say

$$|e^z| \le c_1 \quad \text{and} \quad |h(z)| \le c_2$$

whenever $|z| \le \sqrt{2}$. So for $\beta = \pm\sqrt{2}$, we have

$$|I_\beta| \le |\beta| \, |\beta|^n c_2^{n+1} c_1 \; ;$$

therefore

$$|pqJ| \le 2pqc_1 \left(c_2 \sqrt{2} \right)^{n+1} = c_3 c_4^n \; , \tag{5.14}$$

where c_3 and c_4 are constants which do not depend on n. Now choose n such that $n+1$ is prime, is greater than 2 and $p^2 + q^2$, and is large enough that $n! > c_3 c_4^n$. Then, by the customary arguments, (5.13) shows that pqJ is a non–zero multiple of $n!$, and this contradicts (5.14). Therefore, $e^{\sqrt{2}}$ is irrational.

This proof is rather more involved than the transcendence proof for π, because of all the ps and qs. They arise as a consequence of dealing with an assumed root of the polynomial $qz - p$, whereas previously we were investigating a root of the much simpler polynomial $z + 1$; we alleviated some of the complications by considering only one specific example. We shall do the same in introducing the next difficulty, and ask the reader to contribute by filling in routine details where indicated.

Theorem 5.10. *Exponential of a cubic irrational. If α is a (real or complex) root of the polynomial $z^3 - 3z^2 + 5$, then e^α is irrational.*

Proof. Let α be as stated; write $\alpha_1, \alpha_2, \alpha_3$ for the conjugates of α; suppose that $e^\alpha = p/q$. We have

$$\begin{aligned} 0 &= (qe^{\alpha_1} - p)(qe^{\alpha_2} - p)(qe^{\alpha_3} - p) \\ &= -p^3 + p^2 q(e^{\beta_1} + e^{\beta_2} + e^{\beta_3}) - pq^2(e^{\beta_4} + e^{\beta_5} + e^{\beta_6}) + q^3 e^{\beta_7} \; , \end{aligned} \tag{5.15}$$

where

$$\beta_1 = \alpha_1 \, , \quad \beta_2 = \alpha_2 \, , \quad \beta_3 = \alpha_3 \, ,$$
$$\beta_4 = \alpha_1 + \alpha_2 \, , \quad \beta_5 = \alpha_2 + \alpha_3 \, , \quad \beta_6 = \alpha_3 + \alpha_1 \, ,$$
$$\beta_7 = \alpha_1 + \alpha_2 + \alpha_3 = 3 \, .$$

Now let

$$\begin{aligned} h(z) &= (z - \beta_1)(z - \beta_2) \cdots (z - \beta_7) \\ &= (z - 3)(z^3 - 3z^2 + 5)(z^3 - 6z^2 + 9z - 5) \; ; \end{aligned} \tag{5.16}$$

take $f(z) = z^n h(z)^{n+1}$ and

$$I_\beta = \int_0^\beta f(z) e^{\beta - z} \, dz = F(0) e^\beta - F(\beta) \; , \tag{5.17}$$

where $F = f + f' + f'' + \cdots$. To make effective use of (5.15) we must vary J somewhat from our earlier definition by taking

$$J = p^2 q(I_{\beta_1} + I_{\beta_2} + I_{\beta_3}) - pq^2(I_{\beta_4} + I_{\beta_5} + I_{\beta_6}) + q^3 I_{\beta_7} ; \qquad (5.18)$$

using (5.17) we can then show that

$$\begin{aligned} J = p^3 F(0) &- p^2 q(F(\beta_1) + F(\beta_2) + F(\beta_3)) \\ &+ pq^2(F(\beta_4) + F(\beta_5) + F(\beta_6)) - q^3 F(\beta_7) . \end{aligned} \qquad (5.19)$$

We wish to use properties of symmetric polynomials to prove that the sum of all the terms on the right–hand side, except for the first, is a multiple of $(n + 1)!$. Because of the differing coefficients $p^2 q$, pq^2 and q^3, permuting $\beta_1, \beta_2, \ldots, \beta_7$ does not always leave this expression unchanged, and the symmetry obtaining in our previous arguments has been damaged; but enough remains for us to complete the proof. First, we note that if $j < n + 1$, then $f^{(j)}(\beta_k) = 0$ for every k. Now let $j \geq n + 1$. Then

$$f^{(j)}(x_1) + f^{(j)}(x_2) + f^{(j)}(x_3) \qquad (5.20)$$

is a symmetric polynomial in x_1, x_2, x_3 whose coefficients are multiples of $j!$, and hence of $(n + 1)!$; therefore it can be written as a polynomial P, whose coefficients are also multiples of $(n+1)!$, in the elementary symmetric functions e_1, e_2, e_3. First, we let x_1, x_2, x_3 be $\beta_1, \beta_2, \beta_3$, the roots of $z^3 - 3z^2 + 5$; then the values of e_1, e_2, e_3 are the coefficients of this polynomial, and we have

$$f^{(j)}(\beta_1) + f^{(j)}(\beta_2) + f^{(j)}(\beta_3) = P(e_1, e_2, e_3) = P(3, 0, -5)$$

which is a multiple of $(n+1)!$. Since $\beta_4, \beta_5, \beta_6$ are the roots of $z^3 - 6z^2 + 9z - 5$ we have similarly

$$f^{(j)}(\beta_4) + f^{(j)}(\beta_5) + f^{(j)}(\beta_6) = P(6, 9, 5) ,$$

also a multiple of $(n + 1)!$. Finally,

$$f^{(j)}(\beta_7) = f^{(j)}(3)$$

is a multiple of $(n + 1)!$. Putting all these results together, and evaluating $F(0)$ separately, we can show that

$$J = 75^{n+1} p^3 n! + \{\text{a multiple of } (n + 1)!\} . \qquad (5.21)$$

Let R be a real number greater than the absolute values of all the β_k. Then e^z and $h(z)$ are bounded on $|z| \leq R$, and so we can estimate the integrals appearing in (5.18) to obtain a bound

$$|J| \leq 7|p|^2 |q|^3 R R^n c_2^{n+1} c_1 \leq c_3 c_4^n . \qquad (5.22)$$

If n is chosen suitably, then (5.21) and (5.22) are incompatible; this is a contradiction, and so e^α is irrational.

In exercise 5.5 we ask the reader to fill in the details which have been omitted from this proof.

The additional difficulty referred to in the preamble to Theorem 5.10 is that the expansion (5.15) forces us to consider not only the conjugates $\alpha_1, \alpha_2, \alpha_3$ of the given exponent α but also sums of these conjugates. The success of the proof depended upon two important properties of these sums. Firstly, the expressions fall into three sets

$$\{\beta_1, \beta_2, \beta_3\}, \quad \{\beta_4, \beta_5, \beta_6\}, \quad \{\beta_7\},$$

and each of these sets consists of the roots of some integer polynomial. That is, each is the set of conjugates of some algebraic number. Secondly, for any two β in the same set, the corresponding terms e^β in (5.15) share a common coefficient. This enables us to factor out the coefficient in (5.18) and evaluate J in terms of expressions such as (5.20). Both of these features will be of crucial importance in future proofs. Before turning to the climactic result of this chapter, we give one further proof involving a specific number.

Theorem 5.11. $e^{\sqrt{2}}$ *is transcendental.*

Proof. Suppose, to the contrary, that $e^{\sqrt{2}}$ is an algebraic number of degree m having minimal polynomial

$$p(z) = c_m z^m + c_{m-1} z^{m-1} + \cdots + c_1 z + c_0 \qquad (5.23)$$

over $\mathbb{Z}$, with c_0 and c_m non–zero. Then

$$p(e^{\sqrt{2}})p(e^{-\sqrt{2}}) = 0.$$

Consider the product on the left–hand side,

$$\left(c_0 + c_1 e^{\sqrt{2}} + \cdots + c_m e^{m\sqrt{2}}\right)\left(c_0 + c_1 e^{-\sqrt{2}} + \cdots + c_m e^{-m\sqrt{2}}\right).$$

For every term $c_j e^{j\sqrt{2}} c_k e^{-\sqrt{2}}$ with $j \neq k$ there is another term $c_k e^{k\sqrt{2}} c_j e^{-j\sqrt{2}}$; therefore the terms $e^{(j-k)\sqrt{2}}$ and $e^{-(j-k)\sqrt{2}}$ have the same coefficient and we may write

$$0 = a_0 + a_1\left(e^{\sqrt{2}} + e^{-\sqrt{2}}\right) + \cdots + a_m\left(e^{m\sqrt{2}} + e^{-m\sqrt{2}}\right) \qquad (5.24)$$

with $a_0, a_1, a_2, \ldots, a_m \in \mathbb{Z}$, and specifically

$$a_0 = c_0^2 + c_1^2 + c_2^2 + \cdots + c_m^2 \neq 0.$$

Observe that (5.24) has two properties which have been commented upon above: the sum involves pairs of numbers $\beta, -\beta$, which are conjugates; and the terms $e^{\beta}, e^{-\beta}$ share a common coefficient. The actual values of $a_1, \ldots, a_m$ will turn out to be unimportant; so we have not sought to particularise these coefficients by giving specific formulae, as we did in (5.15). However, it is important – indeed, vital – that the integer term a_0 is not zero; this will not be "automatically" true in future proofs, and will require further argument.

The remainder of the proof will follow familiar lines: once again we shall go through it quite lightly, and invite the reader to fill in details. As in (5.16), we define a monic polynomial with integer coefficients, whose roots are the non–zero exponents in (5.24),

$$h(z) = (z^2 - 2)(z^2 - 8)(z^2 - 18) \cdots (z^2 - 2m^2) \; ;$$

and for any positive integer n we set

$$f(z) = z^n h(z)^{n+1} \; .$$

We shall employ the integrals

$$I_{\beta} = \int_0^{\beta} f(z) e^{\beta - z} \, dz \; ,$$

which are evaluated as

$$I_{\beta} = F(0)e^{\beta} - F(\beta) \quad \text{with} \quad F(z) = f(z) + f'(z) + f''(z) + \cdots \; .$$

Consider

$$J = \sum_{k=1}^{m} a_k \left(I_{k\sqrt{2}} + I_{-k\sqrt{2}} \right)$$

$$= -a_0 F(0) - \sum_{k=1}^{m} a_k \sum_{j=0}^{\infty} \left(f^{(j)}(k\sqrt{2}) + f^{(j)}(-k\sqrt{2}) \right) \; .$$

For any β in $E = \{ \sqrt{2}, -\sqrt{2}, \ldots, m\sqrt{2}, -m\sqrt{2} \}$, the polynomial $f(z)$ has a factor $(z - \beta)^{n+1}$, and so $f^{(j)}(\beta) = 0$ whenever $j \leq n$. If $j \geq n + 1$, then $f^{(j)}(z)$ is a polynomial whose coefficients are integers divisible by $(n+1)!$. So

$$f^{(j)}(z_1) + f^{(j)}(z_2)$$

is a symmetric polynomial in two variables, having coefficients in $(n + 1)! \, \mathbb{Z}$; and $k\sqrt{2}, -k\sqrt{2}$ are the roots of a monic polynomial with integer coefficients; so by Corollary 5.6 on the evaluation of symmetric polynomials,

$$f^{(j)}(k\sqrt{2}) + f^{(j)}(-k\sqrt{2})$$

is a multiple of $(n+1)!$. Evaluating $F(0)$ along the lines of previous proofs holds no surprises, and we obtain

$$J = -a_0 h(0)^{n+1} n! + \{\text{a multiple of } (n+1)!\} \; .$$

Now let c_1 be the maximum absolute value of $a_1, a_2, \ldots, a_m$; set $R = m\sqrt{2}$, so that $|\beta| \leq R$ for all $\beta \in E$. Since $h(z)$ is continuous for all z, there exists c_2 such that $|h(z)| \leq c_2$ whenever $|z| \leq R$; and for similar reasons, the functions $e^{\beta - z}$ with $\beta \in E$ have a common upper bound c_3. Then we have the estimates

$$I_\beta \leq R \cdot R^n c_2^{n+1} c_3$$

for each $\beta \in E$, and

$$|J| \leq 2mc_1(c_2 R)^{n+1} c_3 = c_4 c_5^n \; . \tag{5.25}$$

Finally, choose n such that

$$n! > c_4 c_5^n \quad \text{and} \quad n+1 \text{ is prime} \quad \text{and} \quad n+1 > |a_0|, |h(0)| \; .$$

Then J is not a multiple of $(n+1)!$, so $J \neq 0$; and J is a multiple of $n!$, so

$$|J| \geq n! > c_4 c_5^n \; ,$$

which contradicts (5.25). Therefore, $e^{\sqrt{2}}$ is transcendental.

5.4   TRANSCENDENCE OF e^α

What Lindemann actually proved in 1882 was the following, which has the transcendence of π as an immediate consequence.

Theorem 5.12. Lindemann's Theorem. *If α is a non–zero algebraic number, then e^α is transcendental.*

It is easy to see that this result supersedes all the previous results of this chapter. The proof, however, is a good deal more involved, and so we trust that the reader will not begrudge the time spent on earlier proofs, which have served to introduce many of the fundamental ideas that we shall need. There are still a few issues that we have not yet dealt with, and an informal presentation is likely to be of value; on the other hand, the reader will, no doubt, wish to see a detailed and rigorous proof; we shall give both.

So, let α be algebraic and non–zero; we seek to show that e^α is transcendental. We may assume without loss of generality (*exercise!*) that α is an algebraic integer, and we denote its conjugate algebraic integers by $\alpha_1, \alpha_2, \ldots, \alpha_l$, with $\alpha_1 = \alpha$.

Now suppose that e^{α} is an algebraic number of degree m and has minimal polynomial

$$p(z) = c_m z^m + c_{m-1} z^{m-1} + \cdots + c_1 z + c_0 \qquad (5.26)$$

with integer coefficients, c_m and c_0 being non–zero. Then

$$p\big(e^{\alpha_1}\big)\, p\big(e^{\alpha_2}\big) \cdots p\big(e^{\alpha_l}\big) = 0 , \qquad (5.27)$$

because the first factor is zero. We may use (5.26) to expand the left–hand side of (5.27), giving a sum with integer coefficients which, for the time being, we write "schematically" as

$$\sum_{\beta \in E} \{\text{some coefficient}\}\, e^{\beta} ; \qquad (5.28)$$

our first concern will be to examine this sum more closely. The collection E of exponents consists of all sums of the form

$$\beta = x_1 \alpha_1 + x_2 \alpha_2 + \cdots + x_l \alpha_l$$

with each x_j in $\{0, 1, \ldots, m\}$; note that not all $(m+1)^l$ terms in the collection need be distinct[2]. Our aim will be to define certain integrals I_{β} and a sum something like

$$J = \sum_{\beta \in E} \{\text{some coefficient}\}\, I_{\beta} , \qquad (5.29)$$

– exact details later – and to evaluate the sum in terms of certain function values $F(\beta)$. The procedure should be familiar from earlier proofs. By referring back to these arguments, we see that in general, an individual term $F(\beta)$ could be almost anything, and will give us little information; the way to make progress will be to evaluate sums of the form

$$F(\beta_1) + F(\beta_2) + \cdots + F(\beta_k) , \qquad (5.30)$$

where the β_j are conjugates. To do this by means of symmetry arguments, two things are essential.

- The $\beta_1, \beta_2, \ldots, \beta_k$ in (5.30) must comprise *all* conjugates of some fixed algebraic number. For example, if $\beta_1, \beta_2, \beta_3$ are conjugates, there is not much we can say about an expression such as $F(\beta_1) + F(\beta_2)$.

- The coefficients of terms in (5.28) with conjugate exponents must be the same, so that when evaluating (5.29) we can factor out the common coefficient and leave a "pure" sum such as (5.30).

In terms of the collection E, this comes down to two requirements.

- If any algebraic number occurs in E, then all its conjugates must occur too, and with the same multiplicity.

---

[2] This is why we say "collection" rather than "set", a term which normally indicates that repetitions are to be disregarded.

- If β_1 and β_2 are conjugate elements of E, then the coefficients of e^{β_1} and e^{β_2} in (5.28) must be the same.

These considerations lead to the following definitions.

Definition 5.2. *A* **complete collection of conjugates** *is a finite collection B of algebraic numbers such that if β is in B and $\overline{\beta}$ is an algebraic conjugate of β, then $\overline{\beta}$ is also in B, and occurs with the same multiplicity as β. A* **complete set of conjugates** *is a set consisting of all the conjugates of some algebraic number (occurring once each).*

Examples. The following are complete sets of conjugates, where we write $\zeta = e^{2\pi i/3}$:

- $\{\sqrt{2}, -\sqrt{2}\}$;
- $\{\sqrt[3]{5}, \sqrt[3]{5}\,\zeta, \sqrt[3]{5}\,\zeta^2\}$.

The following are complete collections of conjugates:

- $\{\sqrt{2}, -\sqrt{2}\}$;
- $\{\sqrt{2}, -\sqrt{2}, \sqrt[3]{5}, \sqrt[3]{5}\,\zeta, \sqrt[3]{5}\,\zeta^2\}$;
- $\{\sqrt{2}, -\sqrt{2}, \sqrt{2}, -\sqrt{2}, \sqrt{2}, -\sqrt{2}, \sqrt[3]{5}, \sqrt[3]{5}\,\zeta, \sqrt[3]{5}\,\zeta^2\}$.

The following are not complete collections of conjugates:

- $\{\sqrt{2}, -\sqrt{2}, \sqrt[3]{5}, \sqrt[3]{5}\,\zeta\}$, because $\sqrt[3]{5}\,\zeta^2$ is missing;
- $\{\sqrt{2}, -\sqrt{2}, \sqrt{2}, \sqrt{2}, \sqrt[3]{5}, \sqrt[3]{5}\,\zeta, \sqrt[3]{5}\,\zeta^2\}$, since $\sqrt{2}$ and $-\sqrt{2}$ do not occur the same number of times.

Comment. It is clear that a complete collection of conjugates can be partitioned into subsets which are complete sets of conjugates (and which need not all be distinct).

The following result gives a useful way of identifying complete collections of conjugates.

Lemma 5.13. *A completeness criterion. A collection B of algebraic numbers is a complete collection of conjugates if and only if the polynomial*

$$Q(z) = \prod_{\beta \in B} (z - \beta) \tag{5.31}$$

has rational coefficients. Moreover, if the elements of B are algebraic integers, then $Q(z)$ has integer coefficients.

Proof. Let B be a complete collection of conjugates. Partition B into complete sets of conjugates $B_1, B_2, \ldots, B_k$; each of these consists of the conjugates of some algebraic number β_0, occurring once each. Then

$$\prod_{\beta \in B_j} (z - \beta)$$

is the minimal polynomial of β_0 and therefore has rational coefficients; and $Q(z)$ is the product of these k rational polynomials. Moreover, if B consists of algebraic integers, then each minimal polynomial has integer coefficients, and so does $Q(z)$.

Conversely, suppose that the polynomial $Q(z)$ in (5.31) has rational coefficients, and factorise it into powers of distinct (rational) irreducible polynomials,

$$Q(z) = Q_1(z)^{s_1} Q_2(z)^{s_2} \cdots Q_l(z)^{s_l} .$$

Then any β in B is a root of exactly one $Q_j(z)$; any conjugate $\overline{\beta}$ of β is a root of the same $Q_j(z)$; and β, $\overline{\beta}$ both occur s_j times in B. So B is a complete collection of conjugates.

We shall shortly use this lemma to show that certain subsets of E are complete collections of conjugates. Now we address the issue of the coefficients. The expansion of (5.27) gives a sum of terms

$$c_{x_1} c_{x_2} \cdots c_{x_l} e^{x_1 \alpha_1 + x_2 \alpha_2 + \cdots + x_l \alpha_l}$$

over all l–tuples $\mathbf{x}$ in $X = \{\, 0, 1, \ldots, m \,\}^l$. If $\mathbf{x}$ is in X and $\mathbf{y}$ has the same components as $\mathbf{x}$ but in a different order, then $\mathbf{y}$ is also in X; therefore we can partition X into subsets, each consisting of all possible vectors with a given collection of entries, and this induces a partition of the collection E of exponents in (5.28). It is clear that if $\mathbf{x}$ and $\mathbf{y}$ belong to the same subset of X, then

$$c_{x_1} c_{x_2} \cdots c_{x_l} = c_{y_1} c_{y_2} \cdots c_{y_l} ,$$

and so (5.28) can be written as a sum of sums having the form

$$d_k \sum_{\beta \in D_k} e^\beta , \tag{5.32}$$

where each d_k is a product of certain coefficients from the polynomial (5.26), and the D_k form the partition of E referred to above. We shall show that each D_k is a complete collection of conjugates; therefore the sums (5.32) can be split into sums over complete *sets* of conjugates E_k. It is possible that when this is done, the same complete set of conjugates may occur more than once; but if we collect terms with the same E_k, we find that (5.28) becomes a sum of sums

$$a_k \sum_{\beta \in E_k} e^\beta$$

in which each E_k is a complete set of conjugates, and no two E_k are the same. We note that the coefficient d_k of each sum in (5.32) is an integer; after collecting terms, the coefficients a_k will also be integers.

Example. To illustrate the transformation of the product (5.27) into a sum over complete sets of conjugates, consider the case

$$\alpha_1 = \sqrt{2} + \sqrt{3}, \ \alpha_2 = \sqrt{2} - \sqrt{3}, \ \alpha_3 = -\sqrt{2} + \sqrt{3}, \ \alpha_4 = -\sqrt{2} - \sqrt{3}$$

x	size	β	simplified β
$(0,0,0,0)$	1	0	
$(1,0,0,0)$ etc	4	$\alpha_1, \alpha_2, \alpha_3, \alpha_4$	
$(1,1,0,0)$ etc	6	$\alpha_1 + \alpha_2$ etc	$2\sqrt{2},\, 2\sqrt{3},\, 0,\, 0,$ $-2\sqrt{2},\, -2\sqrt{3}$
$(1,1,1,0)$ etc	4	$\alpha_1 + \alpha_2 + \alpha_3$ etc	$\alpha_1, \alpha_2, \alpha_3, \alpha_4$
$(1,1,1,1)$	1	$\alpha_1 + \alpha_2 + \alpha_3 + \alpha_4$	0

Table 5.1 Sums of conjugates of $\sqrt{2} + \sqrt{3}$.

(so that $l = 4$); and take $m = 1$ (so that we are assuming e^α is rational). We tabulate the quadruples $\mathbf{x}$ in $X = \{0,1\}^4$; the size of the subset of X containing $\mathbf{x}$; the corresponding linear combinations β; and simplified forms of β. The following points are worth noting.

- The β values in lines 2 and 4 are identical, even though they come from different quadruples; the same holds for lines 1 and 5.

- The values in lines $1, 2, 4$ and 5 form complete sets of conjugates.

- The six values in line 3 form a complete collection of conjugates which splits into four complete sets of conjugates

$$\{0\},\ \{0\},\ \{2\sqrt{2}, -2\sqrt{2}\},\ \{2\sqrt{3}, -2\sqrt{3}\};$$

two of these are the same and coincide with the sets in other lines.

Thus, if we write

$$E_0 = \{0\},\ E_1 = \{\alpha_1, \alpha_2, \alpha_3, \alpha_4\},$$
$$E_2 = \{2\sqrt{2}, -2\sqrt{2}\},\ E_3 = \{2\sqrt{3}, -2\sqrt{3}\},$$

then the expansion of (5.27) takes the form

$$a_0 \sum_{\beta \in E_0} e^\beta + a_1 \sum_{\beta \in E_1} e^\beta + a_2 \sum_{\beta \in E_2} e^\beta + a_3 \sum_{\beta \in E_3} e^\beta \tag{5.33}$$

for certain integer coefficients a_0, a_1, a_2, a_3. In this example, it is not hard to bash out the algebra and obtain the explicit expansion

$$(c_0^4 + 2c_0^2 c_1^2 + c_1^4) + (c_0^3 c_1 + c_0 c_1^3)(e^{\alpha_1} + e^{\alpha_2} + e^{\alpha_3} + e^{\alpha_4})$$
$$+ c_0^2 c_1^2 (e^{2\sqrt{2}} + e^{-2\sqrt{2}}) + c_0^2 c_1^2 (e^{2\sqrt{3}} + e^{-2\sqrt{3}}).$$

But the point is that in more complex cases, this computation would be infeasible; so we need to omit the calculation and, instead, understand the general shape of the expression we would have obtained.

As foreshadowed above, we prove a result which will show that the sets D_k in (5.32) are complete collections of conjugates.

Lemma 5.14. *Linear combinations of conjugates. Let $\xi_1, \xi_2, \ldots, \xi_l$ be integers, not necessarily distinct. Let X be the set of all ordered l–tuples $\mathbf{x}$ whose elements are $\xi_1, \xi_2, \ldots, \xi_l$, though not necessarily in that order. Let $\alpha_1, \alpha_2, \ldots, \alpha_l$ be a complete collection of conjugates. Then*

$$E_X = \{\, x_1\alpha_1 + x_2\alpha_2 + \cdots + x_l\alpha_l \mid \mathbf{x} \in X \,\}$$

is a complete collection of conjugates.

Proof. Consider

$$Q(z) = \prod_{\beta \in E_X} (z - \beta) \; :$$

we wish to show that $Q(z)$ is unchanged by any permutation of the α_j. Since any permutation can be composed of successive transpositions of two elements, it suffices to show that E_X, the collection of roots of Q, is unchanged by any transposition of two α_j. Suppose that we interchange α_i and α_j. We can split X into

- a number of individual l–tuples $\mathbf{x}$ in which $x_i = x_j$;

- a number of pairs of vectors

$$\{\, (\ldots, x_i, \ldots, x_j, \ldots), \; (\ldots, x_j, \ldots, x_i, \ldots) \,\}$$

with $x_i \neq x_j$.

The corresponding elements of E_X are

- expressions

$$\cdots + x_i\alpha_i + \cdots + x_j\alpha_j + \cdots$$

which, because $x_i = x_j$, are not altered when α_i and α_j are swapped;

- pairs of expressions

$$\cdots + x_i\alpha_i + \cdots + x_j\alpha_j + \cdots \quad \text{and} \quad \cdots + x_j\alpha_i + \cdots + x_i\alpha_j + \cdots \; ;$$

when α_i and α_j are swapped, each of these expressions becomes the other, so the pair as a whole is unaltered.

Thus E_X is unchanged by any transposition, and therefore by any permutation, of the α_j. Now $Q(z)$ has coefficients which are elementary symmetric polynomials in the elements β of E_X, and hence are polynomials with integer coefficients in the α_j. The argument just given shows that these coefficients are in fact *symmetric* polynomials in the α_j; but as the α_j form a complete collection of conjugates they are the roots of a polynomial with rational coefficients, and so by Corollary 5.6 the coefficients of Q are rational. This completes the proof that E_X is a complete collection of conjugates.

We return to our expansion of (5.27) as a sum of sums over complete sets of conjugates; readers may consider (5.33) as an exemplar. If $E_0 = \{\,0\,\}$, as in this example, we recall from earlier proofs that part of the argument involves showing that the sum J, which will be something like (5.29), is given by an expression of the form

$$J = -a_0 h(0)^{n+1} n! + \{\,\text{a multiple of } (n+1)!\,\}\ ;$$

we then choose n such that, among other requirements, $n+1$ is a prime which is not a factor of a_0. However, if it were to happen that $a_0 = 0$, this would be impossible, and so we would like to rule out this mischance. Unfortunately, in general we can't do so: there are cases in which a_0 turns out to be zero (see, for example, exercise 5.8). We'll then need to modify J in order to make use of a different non–zero coefficient instead of a_0. There are two issues here: firstly, can we be certain that there is a non–zero coefficient in our expansion at all? and secondly, how can we use a non–zero coefficient (other than a_0) to complete the proof?

The first question is not difficult to answer in the affirmative. As a first attempt, we consider a product

$$(s_1 e^{\sigma_1} + \cdots + s_k e^{\sigma_k})(t_1 e^{\tau_1} + \cdots + t_l e^{\tau_l}) \tag{5.34}$$

in which all the σ_j are distinct, all τ_j are distinct, all coefficients s_j and t_j are non–zero, and all σ_j and τ_j are real. By reordering the sums if necessary, we may assume that the smallest of the σ_j is σ_1 and the smallest of the τ_j is τ_1. Then the expansion of (5.34) contains a non–zero term

$$(s_1 t_1) e^{\sigma_1 + \tau_1}\ ; \tag{5.35}$$

any other term in the expansion has exponent larger than $\sigma_1 + \tau_1$. Therefore, when terms having the same exponent are collected, (5.35) remains alone, is not cancelled by any other term, and has a non–zero coefficient.

The defect of the argument just given is that it only deals with real exponents; the exponents α_j that we want for our main proof may well be complex. It turns out that there is less of a difficulty here than might be expected: all we need do is to define what we mean by saying that one complex number is "smaller" than another. This may sound like a doubtful procedure, since it is well known that "it is impossible to order complex numbers"; but this is only true if we want an order which is fully compatible with complex arithmetic, and here we can be satisfied with much less.

Lemma 5.15. Products of exponential sums. *If a product*

$$(s_1 e^{\sigma_1} + \cdots + s_k e^{\sigma_k})(t_1 e^{\tau_1} + \cdots + t_l e^{\tau_l})$$

in which all the σ_j are distinct complex numbers, all τ_j are distinct complex numbers, and all coefficients s_j and t_j are non-zero, is expanded and terms with the same exponent are collected, then there will be at least one term with a non-zero coefficient.

Proof. Define the *lexicographic order* on the set of complex numbers: $z_1 \prec z_2$ means

$$\mathrm{Re}(z_1) < \mathrm{Re}(z_2) \quad \text{or} \quad \mathrm{Re}(z_1) = \mathrm{Re}(z_2) \text{ and } \mathrm{Im}(z_1) < \mathrm{Im}(z_2) \ .$$

Without loss of generality we may assume that σ_1 is the smallest of the σ_j with respect to this ordering, and τ_1 is the smallest of the τ_j. Then the expansion of the product contains a non–zero term

$$(s_1 t_1) e^{\sigma_1 + \tau_1} \ . \tag{5.36}$$

Any other term in the expansion has exponent $\sigma_p + \tau_q$ with either $\sigma_1 \prec \sigma_p$ or $\tau_1 \prec \tau_q$; this exponent is greater than, and therefore not equal to $\sigma_1 + \tau_1$. Hence, when terms having equal exponent are collected, (5.36) remains by itself and has a non–zero coefficient.

Comment. In the above proof we have implicitly used certain "obvious" properties of the lexicographic order which are stated more carefully in the appendix.

Let's review where we are up to. We expand the product (5.27) and collect terms having the same exponent to obtain a sum of the form (5.28), and we know that at least one of the coefficients in this sum is non–zero. Since we have collected terms, no exponent in this sum occurs more than once; we have also shown that terms with conjugate exponents have the same coefficient; and we also know that the exponents can be partitioned into complete sets of conjugates. This means that we have an expression

$$a_0 \sum_{\beta \in E_0} e^\beta + a_1 \sum_{\beta \in E_1} e^\beta + \cdots + a_s \sum_{\beta \in E_s} e^\beta = 0 \tag{5.37}$$

in which every E_k is a complete set of conjugates, no two E_k are the same, and every coefficient a_k is a non–zero integer. If it should happen that $E_0 = \{\,0\,\}$, then we proceed with the kind of argument that we have seen already; but this need not be the case. To obtain a non–zero integer term, we multiply (5.37) by

$$\sum_{\gamma \in E_0} e^{-\gamma} \ . \tag{5.38}$$

We need to ensure that the resulting sum can still be partitioned into sums over complete sets of conjugates, and that it has a non–zero integer term.

Lemma 5.16. *Subtracting complete collections of conjugates. Suppose that B and C are complete collections of conjugates, and write*

$$B - C = \{ \beta - \gamma \mid \beta \in B, \ \gamma \in C \} \, .$$

Then B − C is also a complete collection of conjugates.

Proof. Since B and C are complete collections of conjugates,

$$P_B(z) = \prod_{\beta \in B} (z - \beta) \quad \text{and} \quad P_C(z) = \prod_{\gamma \in C} (z - \gamma)$$

are both polynomials with rational coefficients. Now we can write

$$Q(z) = \prod_{\beta \in B-C} (z - \beta) = \prod_{\gamma \in C} \prod_{\beta \in B} (z - (\beta - \gamma)) = \prod_{\gamma \in C} P_B(z + \gamma) \, ;$$

the coefficients of Q are polynomials with rational coefficients in $\gamma_1, \gamma_2, \ldots, \gamma_t$, the elements of C. Permuting these elements does not alter Q, so the coefficients are symmetric polynomials in the γ_k. Since the γ_k are the roots of P_C, which has rational coefficients, Corollary 5.6 shows that Q has rational coefficients. Thus $B - C$ is a complete collection of conjugates, as claimed.

Now multiply (5.37) by (5.38). We obtain a sum of terms of the form

$$a_k \sum_{\beta \in E_k} \sum_{\gamma \in E_0} e^{\beta - \gamma} = a_k \sum_{\beta \in E_k - E_0} e^{\beta} \, .$$

The lemma guarantees that the right–hand side is a sum over a complete collection of conjugates, and therefore, as we have seen earlier, we can split it into sums over complete sets of conjugates. Moreover, if $k \neq 0$, then E_k and E_0 are distinct complete sets of conjugates and are therefore disjoint, so none of the terms on the right–hand side has zero exponent. The sum over $\beta \in E_0 - E_0$ is split into complete sets of conjugates in the same way; in this case exactly $|E_0|$ of the exponents will be $\beta - \beta = 0$. Hence we have an expression

$$\left(a_0 \sum_{\beta \in E_0} e^{\beta} + a_1 \sum_{\beta \in E_1} e^{\beta} + \cdots + a_s \sum_{\beta \in E_s} e^{\beta} \right) \left(\sum_{\gamma \in E_0} e^{-\gamma} \right)$$

$$= a_0 |E_0| + a_1 \sum_{\beta \in E_1} e^{\beta} + \cdots + a_t \sum_{\beta \in E_t} e^{\beta} = 0 \, ,$$

where, in order not to run out of the letters of the alphabet, we have re–used notation: the sets E_k in the second line need not be the same as the previous E_k, but they are still complete sets of conjugates; and the coefficients a_k need not be the same as the previous a_k, but they are still integers. Moreover, $a_0 \neq 0$, and none of the E_k contains zero.

We have now overcome all the difficulties involved in our generalisation of previous results, and are ready to give a proper proof of the desired theorem.

Theorem 5.12. Lindemann's Theorem [38]. *If α is a non-zero algebraic number, then e^{α} is transcendental.*

Proof. Let α be a non–zero algebraic number; without loss of generality, α is an algebraic integer. Let the conjugates of α be $\alpha_1, \alpha_2, \ldots, \alpha_l$, with $\alpha_1 = \alpha$. Suppose that e^{α} is an algebraic number of degree m having minimal polynomial

$$p(z) = c_m z^m + c_{m-1} z^{m-1} + \cdots + c_1 z + c_0$$

with integer coefficients, where c_m and c_0 are non–zero. We have

$$p(e^{\alpha_1}) \, p(e^{\alpha_2}) \cdots p(e^{\alpha_l}) = 0 \; ,$$

because the first factor is zero. Writing the left–hand side as a product of factors

$$c_0 + c_1 e^{\alpha_k} + c_2 e^{2\alpha_k} + \cdots + c_m e^{m\alpha_k} \tag{5.39}$$

and expanding yields a sum S of terms e^{β} times integer coefficients. If we write

$$X = \{\, 0, 1, 2, \ldots, m \,\}^l$$

for the set of all l–tuples $\mathbf{x} = (x_1, x_2, \ldots, x_l)$ of integers from $\{\, 0, 1, 2, \ldots, m \,\}$, then the exponents β appearing in the sum are the elements of the collection

$$E = \{\, x_1 \alpha_1 + x_2 \alpha_2 + \cdots + x_l \alpha_l \mid \mathbf{x} \in X \,\} \; ;$$

this collection consists of $(m+1)^l$ algebraic integers, not all of which need be distinct. Now partition X into subsets, each consisting of all possible l–tuples with a particular collection of entries, and consider the exponents $\beta \in E$ corresponding to one particular subset X_0. Since $\{\, \alpha_1, \alpha_2, \ldots, \alpha_l \,\}$ is a complete set of conjugates, the "linear combinations of conjugates" lemma shows that these exponents form a complete collection of conjugates, which can be split into one or more complete sets of conjugates. Also, the coefficient of e^{β} is

$$c_{x_1} c_{x_2} \cdots c_{x_l} \; ,$$

and this is the same for all $\mathbf{x} \in X_0$. Therefore, S can be written as a sum of sums

$$c_{\mathbf{x}} \sum_{\beta} e^{\beta} \; ,$$

where in each sum the values of β range over a complete set of conjugates, and each coefficient is an integer. If any two complete sets of conjugates appearing in the sum are the same, we can collect the corresponding terms; thus, we have an expression for S in which no complete set of conjugates appears more than once. Since this expression was obtained by expanding a product of terms like (5.39) and then collecting terms with the same exponent, the "products of exponential sums" lemma, Lemma 5.15, extended inductively to a product of

l factors, guarantees that at least one of the coefficients is non–zero. Choose a term

$$a_0 \sum_{\beta \in E_0} e^\beta$$

with $a_0 \neq 0$, and multiply S by

$$\sum_{\gamma \in E_0} e^{-\gamma} . \tag{5.40}$$

Thanks to the "subtracting complete collections of conjugates" lemma, the new sum still consists of a sum of sums of exponentials over complete collections of conjugates, times integer coefficients; these may once again be split into complete sets of conjugates. The product

$$\sum_{\beta \in E_0} e^\beta \sum_{\gamma \in E_0} e^{-\gamma}$$

contains exactly $|E_0|$ terms $e^{\beta - \beta} = 1$; to put it another way, when $E_0 - E_0$ is decomposed into complete sets of conjugates, $\{0\}$ occurs exactly $|E_0|$ times. Any other product

$$\sum_{\beta \in E_k} e^\beta \sum_{\gamma \in E_0} e^{-\gamma} = \sum_{\beta \in E_k} \sum_{\gamma \in E_0} e^{\beta - \gamma} ,$$

since E_k and E_0 are disjoint, contains no zero exponents. Thus, finally, multiplying S by (5.40) gives an expression

$$a_0 |E_0| + a_1 \sum_{\beta \in E_1} e^\beta + \cdots + a_t \sum_{\beta \in E_t} e^\beta = 0 , \tag{5.41}$$

where the coefficients a_k are integers with $a_0 \neq 0$, and $E_1, E_2, \ldots, E_t$ are complete sets of conjugates not containing zero.

The argument which deduces a contradiction from (5.41) will not differ greatly from earlier proofs in this chapter. For any positive integer n, define

$$h(z) = \prod_{k=1}^{t} \prod_{\beta \in E_k} (z - \beta) , \qquad f(z) = z^n h(z)^{n+1}$$

and

$$I_\beta = \int_0^\beta f(z) e^{\beta - z} \, dz ,$$

where the path of integration is the straight line from 0 to β in the complex plane. We note that the roots of h are algebraic integers forming a complete collection of conjugates, and so $h(z)$ has integer coefficients; moreover, none of the roots is zero, so $h(0) \neq 0$. We can evaluate the integrals as

$$I_\beta = F(0) e^\beta - F(\beta)$$

where

$$F(z) = f(z) + f'(z) + f''(z) + \cdots .$$

Now let

$$J = \sum_{k=1}^{t} \left(a_k \sum_{\beta \in E_k} I_\beta \right) ; \qquad (5.42)$$

we have

$$J = \sum_{k=1}^{t} \left(a_k \sum_{\beta \in E_k} F(0)e^\beta \right) - \sum_{k=1}^{t} \left(a_k \sum_{\beta \in E_k} F(\beta) \right)$$

$$= -a_0 |E_0| F(0) - \sum_{k=1}^{t} a_k \sum_{j=0}^{\infty} \sum_{\beta \in E_k} f^{(j)}(\beta) , \qquad (5.43)$$

noting that the sum over j is really a finite sum, and so there are no convergence issues. Now

$$f^{(j)}(0) = j! \times \{ \text{ coefficient of } z^j \text{ in } f(z) \}$$

$$= \begin{cases} 0 & \text{if } j < n \\ h(0)^{n+1} n! & \text{if } j = n \\ \text{a multiple of } (n+1)! & \text{if } j > n, \end{cases}$$

and so

$$F(0) = h(0)^{n+1} n! + \{ \text{ a multiple of } (n+1)! \} .$$

We need to evaluate the sum involving terms $f^{(j)}(\beta)$. If β is any element of the union of the E_k, then $(z - \beta)^{n+1}$ is a factor of $f(z)$ and so $f^{(j)}(\beta) = 0$ for $j \leq n$. If $j \geq n+1$, then $f^{(j)}(z)$ is a polynomial whose coefficients are integral multiples of $(n+1)!$, and any expression

$$f^{(j)}(z_1) + f^{(j)}(z_2) + \cdots + f^{(j)}(z_s)$$

is a symmetric polynomial in its variables, whose coefficients again are multiples of $(n+1)!$. So for each k, the sum

$$\sum_{\beta \in E_k} f^{(j)}(\beta) \qquad (5.44)$$

is a polynomial with coefficients multiples of $(n+1)!$, evaluated at the elementary symmetric functions of the algebraic integers $\beta \in E_k$. Since E_k is a complete set of conjugates consisting of algebraic integers, each elementary symmetric polynomial is a rational integer, and the sum (5.44) is a multiple of $(n+1)!$. Substituting all these details back into (5.43) yields

$$J = -a_0 |E_0| h(0)^{n+1} n! + \{ \text{ a multiple of } (n+1)! \} . \qquad (5.45)$$

Now we estimate the size of J. The definition of J is a sum of

$$T = |B_1| + |B_2| + \cdots + |B_t|$$

integrals; T here is a fixed finite number, not depending on n. The sum also involves coefficients $a_1, a_2, \ldots, a_t$; let c_0 be the greatest absolute value of these coefficients[3]. Let R be a real number greater than the absolute value of each of the β. The polynomial $h(z)$ is analytic everywhere and therefore bounded on $|z| \leq R$, say $|h(z)| \leq c_1$. The finitely many functions $e^{\beta - z}$ are also analytic everywhere and bounded on $|z| \leq R$; let c_2 be a common bound for all these functions. Therefore

$$|I_\beta| \leq R\,R^n c_1^{n+1} c_2$$

and

$$|J| \leq Tc_0 R^{n+1} c_1^{n+1} c_2 = c_3 c_4^n \ . \tag{5.46}$$

Now choose n such that $n! > c_3 c_4^n$, and such that $n + 1$ is a prime greater than $|a_0|$ and $|E_0|$ and $|h(0)|$. Then (5.45) shows that $|J|$ is an integer which is not a multiple of $(n + 1)!$ and is therefore not zero; but which is a multiple of $n!$ and therefore not less than $n!$; this contradicts (5.46) and completes the proof.

5.5 OTHER RESULTS

The result of the previous section can be rephrased...

Theorem. *Let α_1 and α_2 be unequal algebraic numbers, and let β_1 and β_2 be algebraic numbers. If $\beta_1 e^{\alpha_1} + \beta_2 e^{\alpha_2} = 0$, then $\beta_1 = \beta_2 = 0$.*

...and rephrased again...

Theorem. *Let α_1 and α_2 be unequal algebraic numbers. Then e^{α_1} and e^{α_2} are linearly independent over the field of algebraic numbers.*

...and generalised, giving the extension of Lindemann's result proved by Weierstrass [67] in 1885.

Theorem 5.17. The Lindemann–Weierstrass Theorem. *Let $\alpha_1, \alpha_2, \ldots, \alpha_t$ be unequal algebraic numbers. If $\beta_1, \beta_2, \ldots, \beta_t$ are algebraic numbers and*

$$\beta_1 e^{\alpha_1} + \beta_2 e^{\alpha_2} + \cdots + \beta_t e^{\alpha_t} = 0 \ , \tag{5.47}$$

then $\beta_1 = \beta_2 = \cdots = \beta_t = 0$. That is, $e^{\alpha_1}, e^{\alpha_2}, \ldots, e^{\alpha_t}$ are linearly independent over the field of algebraic numbers.

---

[3]This and subsequent constants need not, naturally, be the coefficients c_k of $p(z)$. But we have (almost if not quite) run out of letters!

Although the details are yet more complicated, the ideas behind the proof of this theorem are no different from those we have already seen. We take (5.47), multiply by similar expressions involving all possible conjugates of the α_k and β_k, and show that we obtain a sum like (5.37). Once again it may be necessary to multiply by an expression like (5.38) in order to ensure that our sum includes a non–zero integer term. We then define a combination of integrals similar to J in (5.42), and obtain contradictory estimates for its divisibility properties and its size.

We conclude this chapter by mentioning two other results in a similar spirit to those listed above. We have already considered results concerning the transcendence of the exponential e^α when α is algebraic; now we let α and β be algebraic and ask whether α^β is algebraic or transcendental. This was the seventh on the list of twenty three problems proposed by David Hilbert in 1900 as a challenge to twentieth–century mathematicians.

It is obvious that α^β is rational if α is 0 or 1; and if β is rational (and α algebraic), then α^β is also algebraic. So we may disregard these cases and consider only $\alpha \neq 0, 1$ and $\beta \notin \mathbb{Q}$. Hilbert, apparently, expected the seventh problem to be one of the most difficult on his list; nevertheless, it was among the first to be solved. Kuzmin, in 1930, showed that α^β is transcendental in the case that β is a quadratic irrational; in 1934, A.O. Gelfond and T. Schneider, working independently of each other, completely settled the problem by proving the following result, and it has become customary to give them equal credit for its discovery.

Theorem 5.18. The Gelfond–Schneider Theorem. *Let α be an algebraic number, not 0 or 1, and β an algebraic irrational. Then α^β is transcendental.*

Note that β may be complex, in which case the power α^β assumes many values; the theorem asserts that *all* of these values are transcendental. The Gelfond–Schneider Theorem has many interesting and simple corollaries.

Corollary 5.19. *The real number e^π is transcendental.*

Proof: e^π is one of the values of i^{-2i}.

Corollary 5.20. Transcendence of logarithms. *If α and γ are algebraic numbers with $\alpha \neq 0, 1$ and $\gamma \neq 0$, then $\log \gamma / \log \alpha$ is either rational or transcendental.*

Proof. If $\beta = \log \gamma / \log \alpha$, then $\alpha^\beta = \gamma$, and so β cannot be an algebraic irrational.

In particular, $\log 2 / \log \frac{3}{2}$, which was of great importance in the musical questions of section 4.4.2, is a transcendental number.

In 1966 the British mathematician Alan Baker proved a series of results which generalise the Gelfond–Schneider Theorem. Baker's main results are generally expressed in terms of the logarithms of algebraic numbers, but in the interests of consistency with the remainder of this chapter we cite two corollaries which can be conveniently written in terms of exponentials. Proofs may be found in [9], Chapter 2.

Theorem 5.21. *If* $\alpha_1, \ldots, \alpha_n, \beta_0, \beta_1, \ldots, \beta_n$ *are non–zero algebraic numbers, then*

$$e^{\beta_0} \alpha_1^{\beta_1} \cdots \alpha_n^{\beta_n}$$

is transcendental.

Theorem 5.22. *If* $\alpha_1, \ldots, \alpha_n$ *are algebraic numbers, not* 0 *or* 1, *and if* $\beta_1, \ldots, \beta_n$ *are algebraic numbers such that* $\beta_1, \ldots, \beta_n$ *and* 1 *are linearly independent over the field of rational numbers, then*

$$\alpha_1^{\beta_1} \cdots \alpha_n^{\beta_n}$$

is transcendental.

Alan Baker
(1939–2018)

Alan Baker was awarded the Fields Medal in 1970.

EXERCISES

5.1 Prove that if a rectangle is inscribed in a circle, then the ratio α of the areas of circle and rectangle, and the ratio β of the perimeters of circle and rectangle, cannot both be algebraic. (From [50].)

5.2 Show that the polynomial

$$(x_1^2 + x_2^2 - x_3^2 - x_4^2)^2 - 4(x_1 x_2 + x_3 x_4)^2$$

is symmetric, and write it as a polynomial in the elementary symmetric polynomials.

5.3 Prove that if $a_1, \ldots, a_n$ are positive constants, then

$$\sum_{k=1}^{n} \frac{1}{1 + a_k} = \frac{e_{n-1} + 2e_{n-2} + 3e_{n-3} + \cdots + ne_0}{(1 + a_1) \cdots (1 + a_n)} \, ,$$

where e_k is the kth elementary symmetric polynomial in $a_1, \ldots, a_n$.

5.4 State and prove a generalisation of exercise 3.16 involving symmetric polynomials.

5.5 Complete the proof of Theorem 5.10 by filling in the following details.

(a) Confirm the equalities (5.16), (5.17), (5.19) and (5.21).

(b) Check that none of the β_k is zero.

(c) Explain why if $j < n+1$, then $f^{(j)}(\beta_k) = 0$.

(d) Show how estimates for the integrals I_{β_k} lead to the bound (5.22).

(e) Explain how n should be chosen in order to obtain a contradiction.

5.6 For $\alpha = \sqrt{2} + \sqrt[3]{3}$, compile a (partial) table of complete collections of conjugates as in the example on page 130, in the case $m = 1$. The minimal polynomial for α was found in exercise 3.1.

(a) What are the conjugates of α?

(b) Find all 15 linear combinations of the α_k corresponding to vectors **x** which are permutations of $(1,1,0,0,0,0)$, and partition these linear combinations into complete *sets* of conjugates. Some of the factorisation and irreducibility testing is rather involved: you may wish to abandon the methods of section 3.1.1 and use computer assistance instead.

(c) Do the same for the 20 permutations of $(1,1,1,0,0,0)$.

5.7 Prove that if e^α is transcendental whenever α is a non–zero algebraic integer, then e^α is transcendental whenever α is a non–zero algebraic number. (This justifies the "without loss of generality" at the beginning of the proof of Theorem 5.12.)

5.8 Let α be the real cube root of 2, and let $\alpha_1, \alpha_2, \alpha_3$ be its conjugates; let $p(z) = 1 - z$. Show that if we expand

$$p(e^{\alpha_1})p(e^{\alpha_2})p(e^{\alpha_3})$$

and collect terms having the same exponent, then the integer term is zero.

5.9 Prove that the following are equivalent:

(1) if α is a non–zero algebraic number, then e^α is transcendental;

(2) if α_1 and α_2 are unequal algebraic numbers, then e^{α_1} and e^{α_2} are linearly independent over the field of algebraic numbers.

5.10 Prove that if α is a non–zero algebraic number, then $\cos\alpha$ and $\sin\alpha$ and $\tan\alpha$ are transcendental.

5.11 Show that there is a unique real c in the interval $(0,1)$ such that

$$\int_0^1 c^x\, dx = \sum_{n=1}^\infty c^n \ .$$

Prove that both c and $\log c$ are transcendental.

5.12 The equation $x^x = 1 + x$ has a unique real positive solution α. Show that α is transcendental.

5.13 Use the Gelfond–Schneider Theorem to prove that if α is a positive algebraic number with $\alpha \neq 1$, then $\cos(\log \alpha)$ and $\sin(\log \alpha)$ are transcendental.

5.14 In Hilbert's statement of his seventh problem, before getting to what is now known as the Gelfond–Schneider Theorem, he gave two versions of a related question.

(1) In an isosceles triangle, if the ratio of the base angle to the vertex angle is algebraic and irrational, then the ratio of the base length to the side length is transcendental.

(2) If β is an algebraic irrational number, then $e^{i\pi\beta}$ is transcendental.

Prove that these two statements are equivalent, and are implied by the Gelfond–Schneider Theorem.

A transcript of Hilbert's address may be found at [32]; there is an English translation at [33].

APPENDIX 1: ROOTS AND COEFFICIENTS OF POLYNOMIALS

Let f be a monic polynomial of degree n given by

$$f(z) = z^n + a_{n-1}z^{n-1} + \cdots + a_1 z + a_0 \, ,$$

and let $\alpha_1, \alpha_2, \ldots, \alpha_n$ be the roots of f, repeated according to multiplicity. Then

$$\alpha_1 + \alpha_2 + \cdots + \alpha_n = -a_{n-1} \quad \text{and} \quad \alpha_1 \alpha_2 \cdots \alpha_n = (-1)^n a_0$$

"and so on".

APPENDIX 2: SOME REAL AND COMPLEX ANALYSIS

Complex integration. The complex contour integral

$$\int_C f(z) \, dz$$

can be defined as a limit of Riemann sums. If the path of integration C is parametrised as

$$z = \phi(t) + i\psi(t) \quad \text{for} \quad a \leq t \leq b \, ,$$

and if we write

$$f(z) = f(x + iy) = u(x, y) + iv(x, y) \, ,$$

then

$$\int_C f(z)\,dz = \int_a^b u(\phi,\psi)\phi' - v(\phi,\psi)\psi'\,dt + i\int_a^b u(\phi,\psi)\psi' + v(\phi,\psi)\phi'\,dt$$

expresses the integral in terms of real integrals. A complex function f is said to be **analytic** at a certain point if it is differentiable throughout some neighbourhood of that point, and **entire** if it is analytic at every point in the complex plane. If f is an entire function (this condition can be relaxed) and if F is another entire function such that $F' = f$, then

$$\int_\alpha^\beta f(z)\,dz = F(\beta) - F(\alpha)$$

is independent of the particular contour from α to β. That is, provided that a complex function is entire, it can be integrated by the same methods as are customarily used for the integration of real functions.

Continuous real and complex functions. Let D be a closed bounded subset of $\mathbb{C}$, and let f be a continuous complex function whose domain includes D. Then f is bounded on D: that is, there exists a constant M such that

$$|f(z)| \le M$$

for all z in D. Moreover, though we don't need the fact for the purposes of this chapter, f actually achieves its maximum: there is a point z_M in D such that

$$|f(z)| \le |f(z_M)|$$

for all z in D.

The above properties are also true for real functions: we can regard these as special cases of complex functions, and $\mathbb{R}$ as a subset of $\mathbb{C}$. In this case D will frequently (though not necessarily) be a closed bounded interval $[a, b]$ on the real line.

Complex logarithms and powers. The logarithm of any non–zero complex number z is defined by

$$\log z = \ln|z| + i\arg(z)\,,$$

where ln denotes the real natural logarithm function; since a given z has many arguments, its logarithm will have many values. If z is a non–zero complex number and β is any complex number, we define the (multi–valued) power

$$z^\beta = \exp(\beta \log z)\,.$$

For example,

$$\log i = \ln|i| + i\arg(i) = i\left(\frac{\pi}{2} + 2k\pi\right)\,, \quad k \in \mathbb{Z}\,;$$

therefore

$$i^{-2i} = \exp\left((-2i)i\left(\frac{\pi}{2} + 2k\pi\right)\right) = \exp(\pi + 4k\pi), \quad k \in \mathbb{Z} ;$$

and one of the values of this expression is e^{π}.

APPENDIX 3: ORDERING COMPLEX NUMBERS

Suppose that we have a set S on which an order is defined. We can use this order to define a related order on S^n, the set of n–tuples of elements from S. We say that $(s_1, \ldots, s_n) \prec (t_1, \ldots, t_n)$, sometimes read as "$(s_1, \ldots, s_n)$ precedes $(t_1, \ldots, t_n)$", if and only if there exists an index k such that

$$s_j = t_j \text{ whenever } j < k, \quad \text{and} \quad s_k < t_k.$$

That is, for the first k at which the n–tuples $\mathbf{s}$ and $\mathbf{t}$ differ, the element of $\mathbf{s}$ is the smaller.

This is known as the (strict) **lexicographic order** on S^n, and a little thought shows that it is really just traditional alphabetical order. To decide which of two words comes first in the dictionary, we compare their first letters; if they are the same we compare the second letters; and so on. We define the lexicographic order on the complex numbers by regarding them as ordered pairs of real numbers: specifically, $z_1 \prec z_2$ if and only if

$$\mathrm{Re}(z_1) < \mathrm{Re}(z_2) \quad \text{or} \quad \mathrm{Re}(z_1) = \mathrm{Re}(z_2) \text{ and } \mathrm{Im}(z_1) < \mathrm{Im}(z_2).$$

We also write $z_1 \preceq z_2$ to mean that either $z_1 \prec z_2$ or $z_1 = z_2$.

Lemma 5.23. Properties of lexicographic order. *The lexicographic order on* $\mathbb{C}$ *has the following properties.*

1. *Trichotomy: for any $z_1, z_2 \in \mathbb{C}$, exactly one of the following is true: $z_1 \prec z_2$ or $z_1 = z_2$ or $z_2 \prec z_1$.*

2. *The order is transitive: for any $z_1, z_2, z_3 \in \mathbb{C}$, if $z_1 \prec z_2$ and $z_2 \prec z_3$, then $z_1 \prec z_3$.*

3. *Any non–empty finite subset A of $\mathbb{C}$ has a smallest element, that is, an element a such that $a \prec z$ for all $z \in A$ other than $z = a$.*

4. *The order is compatible with addition: for all $z_1, z_2, w \in \mathbb{C}$, if $z_1 \prec z_2$ then $z_1 + w \prec z_2 + w$.*

5. *If $z_1 \preceq z_2$ and $w_1 \preceq w_2$ and at least one of the inequalities is strict, then $z_1 + w_1 \prec z_2 + w_2$.*

The proof is left as an exercise.

Comment. Where does this definition leave the well known idea that "it is impossible to order complex numbers"? Well, what is actually meant by this statement is not that it is impossible to order $\mathbb{C}$ at all, but that it is impossible to order $\mathbb{C}$ in a way that is "satisfactory" for most algebraic purposes. Specifically, we want an order $\prec$ which has the properties of trichotomy, transitivity and compatibility with addition already mentioned and is also *compatible with multiplication*: that is, for all $z_1, z_2, w \in \mathbb{C}$,

$$\text{if} \quad z_1 \prec z_2 \text{ and } w \succ 0 , \quad \text{then} \quad z_1 w \prec z_2 w .$$

Exercise. Show by means of an example that the lexicographic order on $\mathbb{C}$ is not compatible with multiplication. This means that the lexicographic order does not conflict with the "impossibility of ordering complex numbers". In fact, it can be proved that there is no way of ordering $\mathbb{C}$ which is "satisfactory" in the sense of the previous comment. Further details may be found in [4].

CHAPTER 6

Automata and Transcendence

A star dies in an exponential arc
And we dribble toward mystery,
Leaving a trail of random decimals.

Tom Petsinis, "A Transcendental Meditation"

IN VIEW OF LIOUVILLE'S THEOREM concerning the poor approximability properties of algebraic numbers, we know that a real number must be transcendental if it is "too well" approximable by rationals. In the examples

$$\alpha = \sum_{k=0}^{\infty} 10^{-2^k} \quad \text{and} \quad \lambda = \sum_{k=1}^{\infty} 10^{-k!}$$

on pages 43 and 50 we found (or attempted to find) good approximations by looking at patterns in the decimal expansions of these numbers and truncating the expansions at suitable points. Similarly, in Chapter 4 we obtained good rational approximations to various numbers – π, the number of days in a year, $\log 2/\log \frac{3}{2}$ – by truncating their continued fractions in an appropriate manner.

These examples suggest that if a number can be expressed in some way (a decimal or a continued fraction; perhaps also an infinite series or product) where the digits or coefficients form some kind of pattern, rather than just a random sequence, then the number "ought to be" transcendental. For by taking advantage of the pattern we may hope to truncate the expression at points which will yield exceptionally good rational approximations; and we know that a number having such approximations must be transcendental.

It is clear, however, that this argument cannot be taken too literally. For a start, it fails for the simplest possible kind of pattern, a periodic sequence, since a decimal with periodic digits is rational and a continued fraction with periodic partial quotients is algebraic (of degree 2). Apart from this, the argument is far too vague to be more than a general guide: many points are

DOI: 10.1201/9781003111207-6

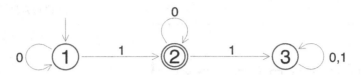

Figure 6.1 A deterministic finite automaton.

left obscure, most importantly the question of what exactly is meant by "a pattern". Defining this term, or, to put it otherwise, distinguishing between "order" and "chaos", is a problem which involves (at least!) mathematics, philosophy and psychology, and which perhaps has no sharply defined answer anyway. To make progress we shall not investigate patterns in general but shall concentrate on producing sequences of numbers by methods sufficiently simple that we may reasonably regard the results as being patterned, though not exhaustive of all possible patterns. We shall then show that, subject to certain technical conditions, these sequences define real numbers which can be proved transcendental.

6.1 DETERMINISTIC FINITE AUTOMATA

A deterministic finite automaton is, roughly speaking, a machine for sorting strings of letters into classes. We can identify a non–negative integer with its string of digits in some base (with the conventions that 0 is represented by the empty string, and that no string begins with zero), and can therefore regard a DFA as a machine for classifying numbers. A simple example of a DFA is shown in figure 6.1. The circles and double circles are called the *states* of the DFA, the numbers inside them being merely labels by which we refer to these states. The arrows between states, marked with numbers, are the *transitions* of the DFA, and the arrow "coming from nowhere" indicates the *initial state*. The transition marked with two numbers and attached to state 3 is really two transitions which have been combined to simplify the diagram. The state marked with a double circle is an *accepting state* while those with single circles are *non–accepting* or *rejecting* states. As far as applications to transcendence are concerned, these ideas should suffice; for readers who care to pursue the topic further, a more formal definition of a DFA is given in the first appendix to this chapter.

Let r be a positive integer and consider a deterministic finite automaton $\mathcal{M}$ over the alphabet $\{0, 1, \ldots, r-1\}$, that is, a DFA whose arrows are marked with these r digits. To see how $\mathcal{M}$ "classifies" a non–negative integer k we write k in base r, start in the initial state and follow the arrows labelled with the digits of k, read from left to right. If after "processing" the last digit we have arrived at an accepting state we say that $\mathcal{M}$ accepts k, if not we say that $\mathcal{M}$ rejects k. The sole function of the DFA (though we may slightly modify this later) is to partition the set of natural numbers into two classes, those accepted and those rejected by $\mathcal{M}$. We may then define the **characteristic sequence** of $\mathcal{M}$ with terms

$$a_k = \begin{cases} 1 & \text{if } \mathcal{M} \text{ accepts } k \\ 0 & \text{if not,} \end{cases}$$

and may write down a "decimal" $a_0.a_1a_2\cdots$ in any base $b \geq 2$. That is, we define

$$\alpha_{\mathcal{M}} = a_0.a_1a_2\cdots = \sum_{k=0}^{\infty} \frac{a_k}{b^k} = \sum_{\mathcal{M} \text{ accepts } k} \frac{1}{b^k} \, .$$

More generally, we regard $\alpha_{\mathcal{M}}$ as being the value at $1/b$ of the function

$$f_{\mathcal{M}}(z) = \sum_{k=0}^{\infty} a_k z^k = \sum_{\mathcal{M} \text{ accepts } k} z^k \, .$$

This is the **generating function** of the sequence $\{a_k\}$, and we shall sometimes also (perhaps imprecisely) refer to it as the generating function of $\mathcal{M}$.

In the example depicted above, we have $r = 2$ and so we classify numbers according to the properties of their binary representation. By convention the first digit of a number is not zero, although this particular automaton has been designed in such a way that the final outcome is no different if it is. It is not difficult to see that in this case a string of zeros and ones leads to and remains in the accepting state, state 2, if and only if it consists of a one followed by a string (possibly empty) of zeros. Therefore, the numbers accepted by the DFA are precisely the powers of 2. The automaton has the generating function

$$f(z) = \sum_{\substack{k \text{ is a} \\ \text{power of } 2}} z^k = \sum_{m=0}^{\infty} z^{2^m} \, , \tag{6.1}$$

and taking, for example, $b = 10$, we may consider the real number

$$\alpha = f\left(\frac{1}{10}\right) = \sum_{m=0}^{\infty} 10^{-2^m} = 0.110100010000000010000\cdots \, .$$

This will be recognised as a number that we have considered as far back as Chapter 1. We noted there that it is easy to see that the decimal expansion of α is not periodic and so α is irrational; in Chapter 4 (exercise 4.10) we showed that α is not approximable to order greater than 2, and therefore is not a Liouville number. We have also asserted that α is transcendental, and we shall soon be able to prove this.

In his autobiographical note *Fifty Years as a Mathematician* [43], Kurt Mahler describes how he began to study transcendence theory in 1926.

During a part of that year I was very ill and in bed. To occupy myself, I played with the function [given above, equation (6.1)] and tried to prove that $f(\zeta)$ is irrational for rational ζ satisfying $0 < |\zeta| < 1$. I succeeded and ended by proving that $f(\zeta)$ is transcendental for all algebraic numbers ζ satisfying this inequality.

We shall study his solution of this particular problem before seeking to generalise the method of proof to further examples.

Comment. Deterministic finite automata are actually extremely weak, and can only recognise the simplest kinds of patterns. (In fact, a DFA is the weakest type of automaton normally regarded as "interesting".) For example, it is possible to prove that there is no DFA which will accept the set of squares $\{0, 1, 4, 9, 16, \dots\}$ and reject the non–squares; so the number

$$\alpha = \sum_{k=0}^{\infty} 10^{-k^2} = 1.1001000010000001000000010000 \cdots$$

from page 7 cannot be produced by the method we have discussed. Roughly, this is because being a square is a "number–theoretical" property, while DFAs can only recognise "typographical" properties. For instance, each of the following sets of natural numbers is accepted by some DFA:

$\{\,$numbers with two consecutive 1s in their base 2 digits$\,\}$,

$\{\,$numbers for which the last 1 is followed by a 2 in base 3$\,\}$,

$\{\,$numbers with an odd number of digits in base 4$\,\}$.

A set which is defined by a number–theoretic property but which *is* accepted by a DFA is the set of all numbers which cannot be written as the sum of three squares,

$$S = \{\, 7, 15, 23, 28, 31, 39, 47, 55, 60, 63, \dots \,\} \,.$$

Exercise. Explain this apparent exception to the "typographical" principle!

6.2 MAHLER'S TRANSCENDENCE PROOF

We begin Mahler's proof by observing some simple facts about the function defined in equation (6.1). First, note that $f(0) = 0$; also, f is defined by a Taylor series convergent for $|z| < 1$ and is therefore analytic in the open unit disc. So we take ζ to be a (complex) algebraic number satisfying $0 < |\zeta| < 1$, assume that $f(\zeta)$ is also algebraic, and seek to obtain a contradiction.

The connection between automata and transcendence relies on obtaining a *functional equation* for the generating function. In the present case, we have

$$f(z) = z + z^2 + z^4 + z^8 + \cdots , \qquad f(z^2) = z^2 + z^4 + z^8 + z^{16} + \cdots$$

and therefore

$$f(z) = f(z^2) + z \; ; \tag{6.2}$$

we shall show that the characteristic sequence of any DFA will lead to either a functional equation of this type, or a system of several such equations in several functions.

The next step is to iterate the functional equation (6.2) – that is, repeatedly substitute the left–hand side into its own right–hand side – to obtain

$$f(z) = f(z^4) + z^2 + z = f(z^8) + z^4 + z^2 + z$$

and in general

$$f(z) = f(z^{2^t}) + z^{2^{t-1}} + \cdots + z^4 + z^2 + z$$

for any $t \geq 0$. The point of this is that then

$$f(\zeta^{2^t}) = f(\zeta) - \zeta^{2^{t-1}} - \cdots - \zeta^4 - \zeta^2 - \zeta \tag{6.3}$$

is an algebraic number which is in a sense no more complicated than ζ and $f(\zeta)$ themselves, but which approaches zero very rapidly as t tends towards infinity. We shall show, however, that a non–zero algebraic number of fixed degree and small denominator cannot be too small in absolute value. This is analogous to the fact that a non–zero rational with denominator q cannot have absolute value less than $1/q$, which we have used in proving various approximation results (see, for example, Lemma 3.18). The inconsistency in the two estimates for the size of our algebraic number will produce the desired contradiction.

In fact, despite the iterated exponential, the expression $f(\zeta^{2^t})$ still does not approach zero fast enough for our purposes, and we must consider the so–called *auxiliary function*

$$E(z) = \sum_{j=0}^{s} a_j(z) f(z)^j \; . \tag{6.4}$$

Here s is a positive integral parameter which, as in proofs by Hermite's method, we shall specify later, choosing it large enough to obtain a contradiction. The expressions $a_j(z)$ are polynomials, not all zero, having integral coefficients and degree at most s; we choose these polynomials in such a way that $E(z)$ has a power series

$$E(z) = \sum_{k=s^2}^{\infty} e_k z^k \tag{6.5}$$

in which the first s^2 coefficients vanish. To see that such a choice for the polynomials is possible, write out the $s + 1$ polynomials $a_j(z)$ in terms of their $(s+1)^2$ unknown coefficients, then expand the sum (6.4) to obtain the first s^2 terms of the power series (6.5); equate the coefficients of all these terms to zero. Then to find the coefficients of the polynomials, we have to solve a homogeneous system of s^2 linear equations in $(s+1)^2$ unknowns, with rational coefficients; but, as is shown in appendix 3, such a system always has a non–zero rational solution. Multiplying by a common denominator gives a non–zero integral solution, as claimed. To clarify this argument we carry out the details with $s = 2$. Set

$$
\begin{aligned}
E(z) &= (a_{00} + a_{01}z + a_{02}z^2) + (a_{10} + a_{11}z + a_{12}z^2)(z + z^2 + z^4 + \cdots) \\
&\quad + (a_{20} + a_{21}z + a_{22}z^2)(z + z^2 + z^4 + \cdots)^2 \\
&= a_{00} + (a_{01} + a_{10})z + (a_{02} + a_{10} + a_{11} + a_{20})z^2 \\
&\quad + (a_{11} + a_{12} + 2a_{20} + a_{21})z^3 + \cdots ;
\end{aligned}
$$

we want to find coefficients a_{jk}, not all zero, in such a way that the terms of degree less than 4 vanish. Since it is sufficient that the coefficients satisfy four homogeneous linear equations in nine unknowns this is certainly possible, and we can find by trial and error a solution such as

$$
a_{00} = a_{01} = a_{10} = a_{12} = a_{20} = a_{22} = 0 \ , \quad a_{02} = a_{21} = 1 \ , \quad a_{11} = -1 \ .
$$

That is, we may choose

$$
E(z) = z^2 - zf(z) + zf(z)^2 \ ;
$$

by substituting the series for $f(z)$ and expanding we can check that the series for $E(z)$ has no terms of degree less than 4.

We must also rule out the possibility that $E(z)$ is identically zero. A function f is said to be **algebraic** over the set $\mathbb{Q}[z]$ of polynomials with rational coefficients if there exist polynomials $a_n, a_{n-1}, \ldots, a_1, a_0$ in $\mathbb{Q}[z]$ such that a_n is not the zero polynomial and

$$
a_n(z)f(z)^n + a_{n-1}(z)f(z)^{n-1} + \cdots + a_1(z)f(z) + a_0(z) = 0 \qquad (6.6)
$$

for all z; the function is said to be **transcendental** over $\mathbb{Q}[z]$ if there are no such polynomials a_j. These definitions are analogous to the definitions of algebraic and transcendental *numbers* that we introduced in Chapter 3.

It can be shown – see appendix 2.4 – that the function f which we are now considering is a transcendental function. Therefore, no equation such as (6.6), with coefficients not all zero, can hold for all z, and it follows from (6.4) that $E(z)$ is not identically zero.

As shown in appendix 2.3, we can estimate the size of $E(z)$ by taking, essentially, its first term alone. Since there exists a complex number β with

$$
|\zeta^{2^t}| < \cdots < |\zeta^2| < |\zeta| < |\beta| < 1 \ ;
$$

and since the series for $f(z)$, and hence also that for $E(z)$, converges at $z = \beta$; we may use the lemma on estimation of power series in the appendix to see that

$$\left| E(\zeta^{2^t}) \right| < c(s) \, |\zeta|^{s^2 2^t} . \tag{6.7}$$

Note that the constant c depends on the function E and therefore on the parameter s – we have indicated this by writing $c(s)$ and not just c – but does not depend on the value $z = \zeta^{2^t}$ at which E is evaluated, and therefore does not depend on t. Observe also that since $|\zeta| < 1$, the right–hand side of (6.7) is a *decreasing* exponential and will be very small for large values of t.

Next we define the *size* of an algebraic number in a way which will serve to measure the number's "algebraic complexity". Recall that for any algebraic number β the denominator, denoted den β, of β is the smallest positive integer d such that $d\beta$ is an algebraic integer, and that if β has degree n, then the conjugates of β are the n complex numbers (including β itself) which have the same minimal polynomial as β.

Definition 6.1. *Let β be an algebraic number with denominator d and conjugates $\beta_1, \beta_2, \ldots, \beta_n$. The* **algebraic size** *of β is*

$$\|\beta\| = \max(d, |\beta_1|, |\beta_2|, \ldots, |\beta_n|) .$$

Note that since d is a positive integer, $\|\beta\|$ is always at least 1.

Examples.

- If $\beta = \sqrt{3} - \sqrt{2}$, then β is an algebraic integer and so its denominator is 1. The conjugates of β are $\pm\sqrt{3} \pm \sqrt{2}$, and the largest of these in absolute value is $\sqrt{3} + \sqrt{2}$. So $\|\beta\| = \sqrt{3} + \sqrt{2}$.

- Let $\beta = \cos \frac{1}{7}\pi$; as we saw in Chapter 3 on pages 33 and 37, the conjugates of β are $\cos \frac{1}{7}\pi$, $\cos \frac{3}{7}\pi$, $\cos \frac{5}{7}\pi$ and its denominator is 2. Since the conjugates of β are less than 1 in absolute value, we have $\|\beta\| = 2$.

- For $\beta = 0$ we have $\|\beta\| = 1$.

The application of this measure to transcendence proofs is based upon the following result.

Lemma 6.1. The fundamental inequality for algebraic size. *If β is a non–zero algebraic number of degree n, then*

$$|\beta| \, \|\beta\|^{2n} \geq 1 .$$

Comment. Roughly speaking, this inequality says that a non–zero algebraic number cannot simultaneously have small denominator and small conjugates, just as – still roughly speaking – a non–zero rational number cannot simultaneously have small denominator and be small itself.

Proof. Let β have denominator d and conjugates $\beta_1, \beta_2, \ldots, \beta_n$. Then $d\beta$ is an algebraic integer whose conjugates are $d\beta_1, d\beta_2, \ldots, d\beta_n$, and so we have a polynomial equation of the form

$$(z - d\beta_1)(z - d\beta_2) \cdots (z - d\beta_n) = z^n + b_{n-1}z^{n-1} + \cdots + b_1 z + b_0 ,$$

where the coefficients b_k on the right–hand side are rational integers. By expanding the left–hand side we obtain

$$|d\beta_1||d\beta_2| \cdots |d\beta_n| = |b_0| ;$$

since none of the conjugates $d\beta_k$ is zero, b_0 cannot be zero and we have

$$1 \leq |b_0| = d^n|\beta_1||\beta_2| \cdots |\beta_n| \leq \|\beta\|^n|\beta|\, \|\beta\|^{n-1} \leq |\beta|\, \|\beta\|^{2n} ,$$

which is the desired result.

Algebraic size has the following properties, which will be proved on page 157.

Lemma 6.2. *Let $\beta_1, \beta_2, \ldots, \beta_m$ be algebraic numbers, and suppose that d is a common denominator for these numbers (that is, $d\beta_j$ is an algebraic integer for every j). Then*

- $\|\beta_1 + \beta_2 + \cdots + \beta_m\| \leq \max(d, \|\beta_1\| + \|\beta_2\| + \cdots + \|\beta_m\|)$;
- $\|\beta_1 + \beta_2 + \cdots + \beta_m\| \leq m\|\beta_1\|\|\beta_2\| \cdots \|\beta_m\|$; *and*
- $\|\beta_1\beta_2 \cdots \beta_m\| \leq \|\beta_1\|\|\beta_2\| \cdots \|\beta_m\|$.

Observe that $E(\zeta^{2^t})$ is an algebraic number for any positive integer t. For since ζ is algebraic and we have assumed that $f(\zeta)$ is also algebraic, relation (6.3) shows that $f(\zeta^{2^t})$ is algebraic; hence

$$E(\zeta^{2^t}) = \sum_{j=0}^{s} a_j(\zeta^{2^t})f(\zeta^{2^t})^j \tag{6.8}$$

is an expression consisting of sums and products of algebraic numbers and is therefore itself algebraic. We can use the above properties to estimate the algebraic size of $E(\zeta^{2^t})$; provided that $E(\zeta^{2^t})$ is not zero, this estimate, together with our estimate (6.7) for its absolute value, will contradict the fundamental inequality for algebraic size and will hence show that $f(\zeta)$ is transcendental.

From (6.8) we see that $E(\zeta^{2^t})$ can be regarded as a polynomial in two variables z_1 and z_2, with degree at most s in each, evaluated at $z_1 = \zeta^{2^t}$ and $z_2 = f(\zeta^{2^t})$. Hence Corollary 6.3 gives the estimate

$$\|E(\zeta^{2^t})\| \leq c'(s)\, \|\zeta\|^{s2^t} \|f(\zeta^{2^t})\|^s .$$

But equation (6.3) expresses $f(\zeta^{2^t})$ as a sum of $t+1$ terms in ζ and $f(\zeta)$, so the properties stated in our most recent lemma give

$$\|f(\zeta^{2^t})\| \le (t+1)\|f(\zeta)\|\|\zeta\|^{2^{t-1}}\cdots\|\zeta\|^4\|\zeta\|^2\|\zeta\| \le (t+1)\|f(\zeta)\|\|\zeta\|^{2^t}$$

and therefore

$$\|E(\zeta^{2^t})\| \le c'(s)(t+1)^s\|f(\zeta)\|^s\|\zeta\|^{2s2^t} . \tag{6.9}$$

We also need to find out something about the degree of $E(\zeta^{2^t})$. To do this we use results on algebraic numbers that we obtained in section 3.1.2. Let V be the vector space over $\mathbb{Q}$ spanned by all terms of the form $\zeta^j f(\zeta)^k$. Then by rewriting the expression (6.8) as

$$E(\zeta^{2^t}) = \sum_{j=0}^{s}\left(a_{j0} + \cdots + a_{js}\zeta^{s2^t}\right)\left(f(\zeta) - \zeta^{2^{t-1}} - \cdots - \zeta^2 - \zeta\right)^j$$

and expanding, we see that $E(\zeta^{2^t})$ is a finite sum of such terms, and hence is an element of V. But V has a spanning set consisting of *finitely many* monomials $\zeta^j f(\zeta)^k$, namely, those for which $0 \le j < \deg\zeta$ and $0 \le k < \deg f(\zeta)$; so V is finite–dimensional, say $\dim V = n$, and we have $\deg E(\zeta^{2^t}) \le n$. Observe that n does not depend on s or t.

Our aim is to show that what we know about the degree, absolute value and algebraic size of $E(\zeta^{2^t})$ contradicts the fundamental inequality, Lemma 6.1. For this inequality to be valid, $E(\zeta^{2^t})$ must be non–zero: we shall use results of complex analysis to prove that this is so for all sufficiently large t. Indeed, if there are infinitely many t with $E(\zeta^{2^t}) = 0$, then $E(z)$ has a sequence of roots tending to the origin, a point at which $E(z)$ is analytic. By the result in appendix 2.3, this implies that $E(z)$ is identically zero. But we already know that this is not so, and therefore $E(\zeta^{2^t})$ is non–zero for all sufficiently large t.

We are now ready to choose s and t in such a way that the estimates (6.7) and (6.9) for the absolute value and algebraic size of $E(\zeta^{2^t})$ are incompatible. To simplify (6.7) write $c_1 = |\zeta|^{-1/2}$, noting that $|\zeta| < 1$ and so $c_1 > 1$. Then increasing powers of c_1 tend to infinity; therefore for any fixed s we can choose t large enough that $c(s) < c_1^{s^2 2^t}$, and consequently

$$|E(\zeta^{2^t})| < c_1^{s^2 2^t} c_1^{-2s^2 2^t} = c_1^{-s^2 2^t} .$$

We use similar ideas to simplify the estimate (6.9): for any s we can find t sufficiently large that $c'(s) < 2^{s 2^t}$, and it is easy to see that $t+1 < 2^{2^t}$ for any $t \ge 0$. Hence

$$\|E(\zeta^{2^t})\| < 2^{s2^t} 2^{s2^t}\|f(\zeta)\|^{s2^t}\|\zeta\|^{2s2^t} = c_2^{s2^t} ,$$

where $c_2 = 4\,\|f(\zeta)\|\|\zeta\|^2$ is a constant independent of s and t. The fundamental inequality for algebraic size now shows that

$$c_1^{-s^2 2^t} c_2^{2ns2^t} \ge 1 , \tag{6.10}$$

provided that $E(\zeta^{2^t}) \neq 0$. Choose an integer $s > 2n \log c_2 / \log c_1$, and then choose t sufficiently large that the simplifications made in this paragraph are valid and $E(\zeta^{2^t})$ is non–zero. Then

$$c_1^{-s^2 2^t} c_2^{2ns2^t} = (c_2^{2n}/c_1^s)^{s2^t} < 1 \, ,$$

which contradicts (6.10) and establishes that $f(\zeta)$ is transcendental.

Although this is rather a long proof, it can be broken up into somewhat less intimidating sections. In summary, we have considered the function

$$f(z) = \sum_{m=0}^{\infty} z^{2^m} = z + z^2 + z^4 + z^8 + \cdots$$

and an algebraic number ζ satisfying $0 < |\zeta| < 1$; we wish to show that $f(\zeta)$ is transcendental by assuming the contrary and deriving a contradiction.

- Find a functional equation for $f(z)$; iterate it, giving a formula for $f(z^{2^t})$.

- Construct an auxiliary function $E(z)$ whose power series begins with a high power of z. The construction involves a parameter s, as yet unspecified.

- For any t, estimate the absolute value of $E(\zeta^{2^t})$.

- Estimate the algebraic size of $E(\zeta^{2^t})$.

- Show that the degree of $E(\zeta^{2^t})$ is independent of s and t.

- Show that for a suitable choice of s and t, the estimates we have made contradict the fundamental inequality for algebraic size; conclude that $f(\zeta)$ cannot be algebraic.

6.3   A MORE GENERAL TRANSCENDENCE RESULT

We seek to generalise the example just investigated. The proof of the generalisation will follow exactly the same lines as that of the example, and we shall set it out according to the steps listed at the end of the previous section. First, we need to prove the properties of algebraic size stated above, as well as various other properties which will facilitate a more general argument.

Lemma 6.2. Properties of algebraic size. *Let $\beta_1, \beta_2, \ldots, \beta_m$ be algebraic numbers, and suppose that d is a common denominator for these numbers, so that $d\beta_j$ is an algebraic integer for all j. Then*

- $\|\beta_1 + \beta_2 + \cdots + \beta_m\| \leq \max(d, \|\beta_1\| + \|\beta_2\| + \cdots + \|\beta_m\|)$;

- $\|\beta_1 + \beta_2 + \cdots + \beta_m\| \leq m\|\beta_1\|\|\beta_2\| \cdots \|\beta_m\|$; *and*

- $\|\beta_1\beta_2 \cdots \beta_m\| \leq \|\beta_1\|\|\beta_2\| \cdots \|\beta_m\|$.

Proof. For any algebraic numbers β and γ, each conjugate of $\beta + \gamma$ has the form $\beta_j + \gamma_k$, where β_j and γ_k are conjugates of β and γ. So if δ is a conjugate of $\beta + \gamma$, then

$$|\delta| = |\beta_j + \gamma_k| \le |\beta_j| + |\gamma_k| \le \|\beta\| + \|\gamma\| \ . \tag{6.11}$$

If d is a common denominator of β and γ, then $d(\beta + \gamma) = d\beta + d\gamma$ is a sum of algebraic integers and so is itself an algebraic integer. Thus the denominator of $\beta + \gamma$ is at most d; this, together with (6.11), shows that

$$\|\beta + \gamma\| \le \max(d, \|\beta\| + \|\gamma\|) \ .$$

The inequality is easily extended by induction to sums of m algebraic numbers, and this proves the first claim. To prove the second, let $d_1, d_2, \ldots, d_m$ be the denominators of $\beta_1, \beta_2, \ldots, \beta_m$ respectively; then it is easy to see that $d_1 d_2 \cdots d_m$ is a common denominator for $\beta_1, \beta_2, \ldots, \beta_m$, and the result we have just proved shows that

$$\|\beta_1 + \beta_2 + \cdots + \beta_m\| \le \max(d_1 d_2 \cdots d_m, \|\beta_1\| + \|\beta_2\| + \cdots + \|\beta_m\|) \ .$$

However

$$d_1 d_2 \cdots d_m \le \|\beta_1\| \|\beta_2\| \cdots \|\beta_m\| \le m\|\beta_1\| \|\beta_2\| \cdots \|\beta_m\| \ ,$$

and since $\|\beta_j\|$ is always at least 1, we have

$$\|\beta_1\| + \|\beta_2\| + \cdots + \|\beta_m\| \le \|\beta_1\| \|\beta_2\| \cdots \|\beta_m\| + \cdots + \|\beta_1\| \|\beta_2\| \cdots \|\beta_m\|$$
$$= m\|\beta_1\| \|\beta_2\| \cdots \|\beta_m\| \ ;$$

so

$$\|\beta_1 + \beta_2 + \cdots + \beta_m\| \le m\|\beta_1\| \|\beta_2\| \cdots \|\beta_m\|$$

as claimed. The proof of the third result is a slightly simpler application of the same ideas, and is left as an exercise.

Corollary 6.3. *Algebraic size of a polynomial expression. Let p be a polynomial in m variables, having degree n_k in the kth variable, with algebraic coefficients. Then there is a constant c, depending only on p, such that for all algebraic numbers $\alpha_1, \alpha_2, \ldots, \alpha_m$, we have*

$$\|p(\alpha_1, \alpha_2, \ldots, \alpha_m)\| \le c \|\alpha_1\|^{n_1} \cdots \|\alpha_m\|^{n_m} \ .$$

Proof. For each k, let d_k be the denominator of α_k; let d be a common denominator for the coefficients of p. Then $D = d \, d_1^{n_1} d_2^{n_2} \cdots d_m^{n_m}$ is a common denominator for all the terms comprising $p(\alpha_1, \alpha_2, \ldots, \alpha_m)$; clearly

$$D \le d \|\alpha_1\|^{n_1} \cdots \|\alpha_m\|^{n_m} \ .$$

Using the third of the above properties, each term in $p(\alpha_1, \alpha_2, \ldots, \alpha_m)$ has algebraic size

$$\|p_{j_1 \cdots j_m} \alpha_1^{j_1} \cdots \alpha_m^{j_m}\| \leq \|p_{j_1 \cdots j_m}\| \, \|\alpha_1\|^{j_1} \cdots \|\alpha_m\|^{j_m}$$
$$\leq \|p_{j_1 \cdots j_m}\| \, \|\alpha_1\|^{n_1} \cdots \|\alpha_m\|^{n_m} \, ,$$

and the sum of the algebraic sizes of all terms is at most

$$\left(\sum_{j_1, \ldots, j_m} \|p_{j_1 \cdots j_m}\| \right) \|\alpha_1\|^{n_1} \cdots \|\alpha_m\|^{n_m} = P \, \|\alpha_1\|^{n_1} \cdots \|\alpha_m\|^{n_m} \, ,$$

say. So by the first property,

$$\|p(\alpha_1, \alpha_2, \ldots, \alpha_m)\| \leq \max\left(d \, \|\alpha_1\|^{n_1} \cdots \|\alpha_m\|^{n_m}, \, P \, \|\alpha_1\|^{n_1} \cdots \|\alpha_m\|^{n_m} \right)$$
$$= c \, \|\alpha_1\|^{n_1} \cdots \|\alpha_m\|^{n_m} \, ,$$

where $c = \max(d, P)$ is a constant which depends only on the polynomial p.

Next we prove a result concerning products of polynomials evaluated at powers of a fixed algebraic number. This will be required in proving Theorem 6.6, which generalises our earlier example.

Corollary 6.4. Algebraic size of a polynomial product. *Let $p_1, p_2, p_3, \ldots, p_t$ be polynomials with integer coefficients, having degree less than m and height less than h. Let ζ be a non-zero algebraic number, and let r be an integer, $r \geq 2$. Then*

$$\left\| p_1(\zeta) \, p_2(\zeta^r) \, p_3(\zeta^{r^2}) \cdots p_t(\zeta^{r^{t-1}}) \right\| < m^t h^t \|\zeta\|^{mr^t} \, .$$

Proof. The expression

$$p_1(\zeta) \, p_2(\zeta^r) \, p_3(\zeta^{r^2}) \cdots p_t(\zeta^{r^{t-1}}) \tag{6.12}$$

is a sum of at most m^t terms $c\zeta^k$, where each coefficient c is an integer with $|c| < h^t$, and the degree satisfies

$$k < m + mr + mr^2 + \cdots + mr^{t-1} < mr^t \, .$$

Each term has algebraic size at most $|c| \, \|\zeta\|^k < h^t \|\zeta\|^{mr^t}$; and all these terms have a common denominator $(\text{den} \, \zeta)^{mr^t}$, which is therefore a denominator for 6.12. Noting that this denominator is at most $\|\zeta\|^{mr^t}$ and using the first of the properties of algebraic size in Lemma 6.2, we have

$$\left\| p_1(\zeta) \, p_2(\zeta^r) \, p_3(\zeta^{r^2}) \cdots p_t(\zeta^{r^{t-1}}) \right\| < \max\left(\|\zeta\|^{mr^t}, \, m^t h^t \|\zeta\|^{mr^t} \right) \, ,$$

and the result follows.

The following result has already been proved on page 153; we restate it here for convenience.

Lemma 6.1. The fundamental inequality for algebraic size. *If β is a non–zero algebraic number of degree n, then $|\beta| \, \|\beta\|^{2n} \geq 1$.*

Finally, we shall need an estimate for the size of the reciprocal of a non–zero algebraic number.

Lemma 6.5. Algebraic size of a reciprocal. *If β is a non-zero algebraic number of degree n, then $\|\beta^{-1}\| \leq \|\beta\|^{2n}$.*

Proof. Suppose that β is a non–zero root of the irreducible polynomial

$$b_n z^n + b_{n-1} z^{n-1} + \cdots + b_1 z + b_0 \qquad (6.13)$$

whose coefficients are rational integers with no common factor. Then β^{-1} is a root of

$$b_0 z^n + b_1 z^{n-1} + \cdots + b_{n-1} z + b_n , \qquad (6.14)$$

which is also irreducible. Consequently, the conjugates of β^{-1} are the reciprocals of the conjugates $\beta_1, \beta_2, \ldots, \beta_n$ of β. From the fundamental inequality, we have

$$|\beta_j^{-1}| \leq \|\beta_j\|^{2n} = \|\beta\|^{2n} .$$

Now let d be the denominator of β. Using (6.14) and then (6.13), we have

$$\mathrm{den}(\beta^{-1}) \leq |b_0| = |b_n| |\beta_1| |\beta_2| \cdots |\beta_n| \leq |b_n| \|\beta\|^n .$$

On the other hand, $d\beta$ is an algebraic integer, so there is a polynomial equation

$$d^n \beta^n + c_{n-1} d^{n-1} \beta^{n-1} + \cdots + c_1 d\beta + c_0 = 0 ,$$

where the coefficients c_k are rational integers. Comparing this with (6.13), in which the b_k have no common factor, we see that $|b_n| \leq d^n$; hence

$$\mathrm{den}(\beta^{-1}) \leq d^n \|\beta\|^n \leq \|\beta\|^{2n} .$$

Combining the estimates for the denominator and conjugates of β^{-1} establishes the lemma.

All these results on algebraic size enable us to prove a generalisation of Mahler's original example.

Theorem 6.6. Functional equations and transcendence. *Let f be a function having a Taylor series with integer coefficients, convergent inside the unit circle. Suppose that f satisfies a functional equation*

$$f(z) = a(z) f(z^r) + b(z) ,$$

where $r \geq 2$ is an integer, $a(z)$ and $b(z)$ are polynomials with integral coefficients, and $a(z)$ has no roots inside the unit circle, except possibly at $z = 0$. Let ζ be an algebraic number with $0 < |\zeta| < 1$. If f is a transcendental function, then $f(\zeta)$ is a transcendental number.

Proof. *Step 1: iterating the functional equation.* We have

$$f(z) = a(z)f(z^r) + b(z)$$
$$= a(z)a(z^r)f(z^{r^2}) + a(z)b(z^r) + b(z)$$
$$= a(z)a(z^r)a(z^{r^2})f(z^{r^3}) + a(z)a(z^r)b(z^{r^2}) + a(z)b(z^r) + b(z)$$
$$= \cdots$$
$$= A_t(z)f(z^{r^t}) + B_t(z) ,$$

where

$$A_t(z) = a(z)a(z^r)a(z^{r^2})\cdots a(z^{r^{t-1}})$$

and

$$B_t(z) = b(z) + a(z)b(z^r) + \cdots + a(z)a(z^r)\cdots a(z^{r^{t-2}})b(z^{r^{t-1}}) .$$

Step 2: construction of the auxiliary function. For any positive integer s there exist polynomials $p_0(z), p_1(z), \ldots, p_s(z)$ with integral coefficients, not all zero, having degree at most s, such that there is a power series

$$E(z) = \sum_{j=0}^{s} p_j(z)f(z)^j = \sum_{k=s^2}^{\infty} e_k z^k .$$

This is true because it requires that the $(s+1)^2$ coefficients of the polynomials satisfy a system of s^2 homogeneous equations with integral coefficients; since there are more unknowns than equations this system certainly has a non-trivial rational solution, and multiplying by a common denominator gives an integral solution. Moreover, $E(z)$ is not identically zero because $f(z)$ is a transcendental function.

Step 3: estimation of absolute value. Since $f(z)$ is analytic inside the unit circle, so is $E(z)$. From the estimate (6.20) in appendix 2.3, we have

$$\left|E\left(\zeta^{r^t}\right)\right| < c(s)\,|\zeta|^{s^2 r^t} ,$$

where the constant $c(s)$ does not depend on t. Let $c_1 = |\zeta|^{-1/2}$. Then $c_1 > 1$, and the above inequality can be written

$$\left|E\left(\zeta^{r^t}\right)\right| < c(s)\,c_1^{-2s^2 r^t} .$$

Note that the constant c_1 is independent of both s and t; the same will be true of the constants $c_2, c_3, \ldots$ which will arise in the course of the proof.

Step 4: estimation of degrees. From the functional equation we have

$$f\left(\zeta^{r^t}\right) = \frac{f(\zeta) - B_t(\zeta)}{A_t(\zeta)} ,$$

noting that the denominator is not zero since $a(z)$ is not zero at any of the points ζ^{r^k}. Hence $f(\zeta^{r^t})$ is an element of the vector space over $\mathbb{Q}$ spanned by products of powers of ζ and powers of $f(\zeta)$, and so too is

$$E(\zeta^{r^t}) = \sum_{j=0}^{s} p_j(\zeta^{r^t}) f(\zeta^{r^t})^j .$$

Since by assumption both ζ and $f(\zeta)$ are algebraic, this space has (finite) dimension n, independent of s and t.

Step 5: estimation of algebraic size. We can regard $E(\zeta^{r^t})$ as a polynomial in two variables z_1, z_2, having integral coefficients and degree at most s in each, evaluated at $z_1 = \zeta^{r^t}$ and $z_2 = f(\zeta^{r^t})$. So by Corollary 6.3, we have

$$\|E(\zeta^{r^t})\| \le c'(s) \|\zeta^{r^t}\|^s \|f(\zeta^{r^t})\|^s .$$

Now suppose that a and b have degrees less than m and heights less than h; note that these numbers are independent of the parameters s and t. Then $A_t(\zeta)$ is an expression of the type considered in Corollary 6.4, and $B_t(\zeta)$ is a sum of t such expressions having a common denominator at most $(\text{den } \zeta)^{mr^t}$; therefore

$$\|A_t(\zeta)\| < m^t h^t \|\zeta\|^{mr^t} < c_2^{r^t} \quad \text{and} \quad \|B_t(\zeta)\| < t m^t h^t \|\zeta\|^{mr^t} < (2c_2)^{r^t} ,$$

where $c_2 = mh\|\zeta\|^m$ is a constant which does not depend on s or t. The lemma concerning the algebraic size of a reciprocal then yields

$$\|f(\zeta^{r^t})\| = \left\| \frac{f(\zeta) - B_t(\zeta)}{A_t(\zeta)} \right\|$$
$$\le \|f(\zeta) - B_t(\zeta)\| \|A_t(\zeta)\|^{2n}$$
$$< 2\|f(\zeta)\| (2c_2)^{r^t} c_2^{2nr^t}$$
$$< c_3^{r^t}$$

with $c_3 = 4\|f(\zeta)\| c_2^{2n+1}$, and so

$$\|E(\zeta^{r^t})\| < c'(s) \|\zeta\|^{sr^t} c_3^{sr^t} = c'(s) c_4^{sr^t} .$$

Step 6: conclusion. Choose s such that $c_4^{4n} < c_1^s$; this can be done since $c_1 > 1$ and c_1, c_4 and n do not depend on s. Then we can choose t with the following properties:

- $c(s) < c_1^{s^2 r^t}$ and $c'(s) < c_4^{s r^t}$: this, again, is possible because $c(s), c'(s),$ c_1 and c_4 do not depend on t;

- $E(\zeta^{r^t}) \ne 0$: this is possible because $E(z)$ is analytic and not identically zero, and therefore cannot have a sequence of roots converging to zero (see appendix 2.2).

Finally, the fundamental inequality for the size of a non–zero algebraic number gives

$$|E(\zeta^{r^t})|\,\|E(\zeta^{r^t})\|^{2n} \geq 1$$

while from the above estimates, we have

$$|E(\zeta^{r^t})|\,\|E(\zeta^{r^t})\|^{2n} < c(s)c_1^{-2s^2r^t}c'(s)^{2n}c_4^{2nsr^t}$$

$$< c_1^{s^2r^t}c_1^{-2s^2r^t}c_4^{2nsr^t}c_4^{2nsr^t} = \left(\frac{c_4^{4n}}{c_1^s}\right)^{sr^t} < 1\,.$$

We have a contradiction, and the result is proved.

Comment. In order to simplify the proof, we have assumed more than we really need to in the statement of Theorem 6.6. For example, the only reason for the stipulation that $a(z)$ has no roots inside the unit circle was to ensure that

$$A_t(\zeta) = a(\zeta)a(\zeta^r)a(\zeta^{r^2})\cdots a(\zeta^{r^{t-1}})$$

is not zero; clearly it would have been sufficient to demand that

$$a(\zeta^{r^k}) \neq 0 \quad \text{for} \quad k = 0, 1, 2, 3, \dots \,.$$

Nor is it essential that $a(z)$ and $b(z)$ be polynomials; if they are rational functions with integral coefficients the proof works in much the same way, though the estimates involving A_t and B_t become more intricate. Finally, the requirement that the coefficients of $a(z)$ and $b(z)$, and the Taylor series coefficients of $f(z)$, be integers is also more stringent than necessary; it suffices that the field generated by all these coefficients be of finite degree when considered as a vector space over $\mathbb{Q}$.

6.4 A TRANSCENDENCE PROOF FOR THE THUE SEQUENCE

We can apply similar methods to show that a "decimal" whose digits comprise the Thue sequence is transcendental. In Chapter 1 the sequence was defined recusively thus,

$$a_{2k} = a_k\,, \quad a_{2k+1} = 1 - a_k = \begin{cases} 1 & \text{if } a_k = 0 \\ 0 & \text{if } a_k = 1, \end{cases}$$

with the initial condition $a_0 = 0$. Using the characterisation (page 8) of a_k as the parity of the binary expansion of k, we see that the Thue sequence is recognised by the DFA $\mathcal{M}$ shown in figure 6.2. We can write down the generating function of $\mathcal{M}$ and rearrange the infinite series,

$$f(z) = \sum_{k=0}^{\infty} a_k z^k = \sum_{k=0}^{\infty} a_{2k}z^{2k} + \sum_{k=0}^{\infty} a_{2k+1}z^{2k+1}$$

$$= \sum_{k=0}^{\infty} a_k z^{2k} + \sum_{k=0}^{\infty}(1 - a_k)z^{2k+1} = (1 - z)\sum_{k=0}^{\infty} a_k z^{2k} + \sum_{k=0}^{\infty} z^{2k+1}$$

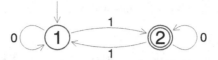

Figure 6.2 A deterministic finite automaton for the Thue sequence.

to obtain the functional equation

$$f(z) = (1 - z)f(z^2) + \frac{z}{1 - z^2} \cdot \quad (6.15)$$

This functional equation does not fall within the scope of Theorem 6.6 since one of the terms involves a rational function which is not a polynomial. As mentioned above, this difficulty is not insuperable, and the theorem could be extended so as to cover this case; however, instead of doing so we can prove an irrationality result for $f(\zeta)$ by considering a closely related function.

Let $\{\, b_k \,\}$ be the Thue sequence on $\{\, 1, -1 \,\}$, that is, the sequence which is obtained from the original Thue sequence when every 0 is replaced by 1 and every 1 by -1. It is not hard to see that $b_0 = 1$, that the sequence satisfies the recurrence

$$b_{2k} = b_k \,, \quad b_{2k+1} = -b_k$$

for $k \geq 0$, and that

$$b_k = 1 - 2a_k$$

for every k. The generating function $g(z)$ for $\{\, b_k \,\}$ is related to the generating function $f(z)$ for $\{\, a_k \,\}$ by

$$g(z) = \sum_{k=0}^{\infty} b_k z^k = \sum_{k=0}^{\infty} (1 - 2a_k) z^k$$

$$= \sum_{k=0}^{\infty} z^k - 2 \sum_{k=0}^{\infty} a_k z^k = \frac{1}{1 - z} - 2f(z) \,. \quad (6.16)$$

If ζ is algebraic and $0 < |\zeta| < 1$, then this identity shows that $g(\zeta)$ is algebraic if and only if $f(\zeta)$ is algebraic. There is also a functional equation for $g(z)$: we can use (6.16) and the known functional equation for $f(z)$, or we can start from scratch to find

$$g(z) = \sum_{k=0}^{\infty} b_k z^k = \sum_{k=0}^{\infty} b_{2k} z^{2k} + \sum_{k=0}^{\infty} b_{2k+1} z^{2k+1}$$

$$= \sum_{k=0}^{\infty} b_k z^{2k} - \sum_{k=0}^{\infty} b_k z^{2k+1} = (1 - z)g(z^2) \,.$$

Incidentally, this shows that $g(z)$ has an elegant representation as an infinite product,

$$g(z) = (1-z)(1-z^2)(1-z^4)(1-z^8)\cdots = \prod_{k=0}^{\infty}\left(1-z^{2^k}\right) \ .$$

Exercise. Use a combinatorial argument, together with the fact that a_k is the parity of the binary representation of k, to prove the same result.

It is now easy to show that if ζ is algebraic and $0 < |\zeta| < 1$, then $g(\zeta)$ is transcendental. The conditions of our main theorem are satisfied by

$$r = 2 \ , \quad a(z) = 1 - z \quad \text{and} \quad b(z) = 0 \ ;$$

and using the results quoted from Pólya and Szegő in appendix 2.4, we see that g is a transcendental function; so $g(\zeta)$ is transcendental. It follows from previous remarks that $f(\zeta)$ is also transcendental, and a particular corollary is that the decimal

$$\tau = f\left(\frac{1}{10}\right) = 0.110100110010110100010110\cdots \ ,$$

which we proved irrational in Chapter 1, is in fact transcendental.

To employ the method of the present chapter it is not necessary to have a generating function in the form of an infinite series or product: the important thing is to be able to write down a suitable functional equation. For example, we could speculatively define f by the continued fraction

$$f(z) = z + \cfrac{1}{z^3 +} \ \cfrac{1}{z^9 +} \ \cfrac{1}{z^{27} +} \ \cdots$$

for $|z| > 1$; then f satisfies the relation

$$f(z) = z + \frac{1}{f(z^3)} \ ,$$

which is somewhat similar to the functional equation (6.2), and we might attempt to use this equation as the basis of a transcendence proof. It is sometimes necessary, as we shall see in the next few pages, to consider not single functions but n-tuples of functions, satisfying functional equations in which the coefficients are matrices of rational functions.

Mahler's work from the 1920s and 1930s was largely forgotten at the time but was revived and greatly extended by Loxton and van der Poorten, Kubota, Nishioka, and others, during the 1970s and 1980s. It serves as a fascinating connection between the topics of automata, functional equations and transcendence theory.

Kurt Mahler
(1903–1988)

6.5 AUTOMATA AND FUNCTIONAL EQUATIONS

To conclude this chapter we investigate the relation between deterministic finite automata and functional equations. Recall that at the beginning of this chapter we calculated the set of numbers accepted by a given DFA, wrote down its generating function and by means of simple algebra found a functional equation satisfied by the function. Later on (section 6.4) we found an equation for the generating function of the Thue sequence by manipulating power series and using the recurrence relation which defines the sequence. It is possible, however, to derive a functional equation, or a system of such equations, directly from a DFA, thus demonstrating the close connection between automata and functional equations.

Let $r \geq 2$; we shall understand that when a natural number is written in base r, the first digit is never zero (and, in particular, 0 is represented by the empty string). Let $\mathcal{M}$ be a DFA over the alphabet $\{0, 1, \ldots, r-1\}$ and suppose that $\mathcal{M}$ has s states, numbered $1, 2, \ldots, s$. For each state m and each non–negative integer k write $f_{m,k} = 1$ if the base r digits of k, read as usual from left to right, lead from the initial state of $\mathcal{M}$ to state m; and $f_{m,k} = 0$ otherwise. We define s generating functions

$$f_m(z) = \sum_{k=0}^{\infty} f_{m,k} z^k \; ;$$

that is, $f_m(z)$ is the sum of z^k for all k which end up in state m when written in base r and "processed" by the automaton. We seek relations between the functions $f_m(z)$ and $f_m(z^r)$.

We shall begin by splitting up the sum defining $f_m(z)$ in much the same way as we did on page 162. Dividing a non–negative integer by r to give quotient k and remainder j, we have

$$f_m(z) = \sum_{j=0}^{r-1} \sum_{k=0}^{\infty} f_{m,rk+j} z^{rk+j} \; ,$$

and we now need to determine the connection between the coefficients $f_{m,rk+j}$ and $f_{m,k}$. The digits of $rk + j$ are those of k, with a digit j appended at the right–hand end, and so $\mathcal{M}$ takes $rk + j$ to state m if and only if it takes k to some state m' from which an arrow marked j leads to m. Symbolically,

$$f_{m,rk+j} = 1 \quad \text{if and only if} \quad f_{m',k} = 1 \text{ and } m' \xrightarrow{j} m \text{ for some } m' \; ,$$

and this may be restated as

$$f_{m,rk+j} = 1 \quad \text{if and only if} \quad f_{m_1,k} = 1 \text{ or } f_{m_2,k} = 1 \text{ or } \ldots \text{ or } f_{m_t,k} = 1 \; ,$$

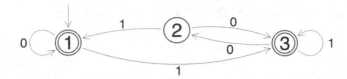

Figure 6.3 A deterministic finite automaton.

where $m_1, m_2, \ldots, m_t$ are the states from which an arrow marked j leads to m. (Observe, incidentally, that m itself may be one of these states.) To simplify this result note that

$$0 \le f_{m_1,k} + f_{m_2,k} + \cdots + f_{m_t,k} \le \sum_{\text{all } m} f_{m,k} = 1 ,$$

the final equality being true because the prohibition on leading zeros means that the "processing" of k leads to one and only one of the states of $\mathcal{M}$. We also know that each individual $f_{m,k}$ is either 0 or 1, and hence the last equivalence can be expressed as

$$f_{m,rk+j} = 1 \quad \text{if and only if} \quad f_{m_1,k} + f_{m_2,k} + \cdots + f_{m_t,k} = 1 ,$$

or simply

$$f_{m,rk+j} = f_{m_1,k} + f_{m_2,k} + \cdots + f_{m_t,k} . \tag{6.17}$$

Before proceeding further we illustrate with an example.

For the automaton shown in figure 6.3, we have $r = 2$; and we shall first consider state 2. The only arrow leading to state 2 is that which originates at state 3 and is marked 0. An odd integer, written in base 2, ends with a 1 and so cannot be taken to state 2 by the DFA; thus $f_{2,2k+1} = 0$ for all k. An even integer ends in 0, and the DFA takes $2k$ to state 2 if and only if it takes k to state 3; so $f_{2,2k} = 1$ if and only if $f_{3,k} = 1$. Since the $f_{m,k}$ can only take the values 0 and 1, this is equivalent to saying that $f_{2,2k} = f_{3,k}$.

Consider state 3. By similar arguments to those above, we have $f_{3,2k} = f_{2,k}$. To evaluate $f_{3,2k+1}$ we must note that *two* arrows marked 1 point to state 3; thus the DFA takes $2k+1$ to state 3 if and only if it takes k either to state 1 or to state 3. We have

$$f_{3,2k+1} = f_{1,k} + f_{3,k} .$$

To see this remember that each of the three terms is either 0 or 1. If $f_{3,2k+1} = 1$, then processing k through $\mathcal{M}$ leads to state 1 or 3, so either $f_{1,k}$ or $f_{3,k}$ must be 1, and as they cannot both be 1 their sum is 1; while if $f_{3,2k+1} = 0$, then processing k leads to neither of these states, so $f_{1,k} = f_{3,k} = 0$.

Continuing the general argument, we may use the recurrence (6.17) to find a system of functional equations for the $f_m(z)$. We have

$$f_m(z) = \sum_{j=0}^{r-1} \sum_{k=0}^{\infty} f_{m,rk+j} z^{rk+j} = \sum_{j=0}^{r-1} \sum_{m' \xrightarrow{j} m} \sum_{k=0}^{\infty} f_{m',k} z^{rk+j}$$

$$= \sum_{j=0}^{r-1} \sum_{m' \xrightarrow{j} m} z^j \sum_{k=0}^{\infty} f_{m',k} z^{rk} = \sum_{j=0}^{r-1} \sum_{m' \xrightarrow{j} m} z^j f_{m'}(z^r) ;$$

here the inner sum is over all states m' from which an arrow labelled j goes to state m. This identity may be written as

$$f_m(z) = \sum_{m'=1}^{s} w_{m,m'}(z) f_{m'}(z^r) , \qquad (6.18)$$

where

$$w_{m,m'}(z) = \sum_{j} z^j ,$$

the sum being over all j for which the DFA $\mathcal{M}$ has an arrow marked j pointing from state m' to state m. In fact, we may define a vector and a matrix

$$f(z) = \begin{pmatrix} f_1(z) \\ \vdots \\ f_s(z) \end{pmatrix} \quad \text{and} \quad W(z) = \begin{pmatrix} w_{1,1}(z) & \cdots & w_{1,s}(z) \\ \vdots & \ddots & \vdots \\ w_{s,1}(z) & \cdots & w_{s,s}(z) \end{pmatrix},$$

whereupon the system of s functional equations becomes the matrix–vector equation

$$f(z) = W(z) f(z^r) . \qquad (6.19)$$

Examples.

- Consider the DFA in figure 6.2, which recognises the Thue sequence. In this case $s = 2$ and (6.18) becomes

$$f_1(z) = f_1(z^2) + z f_2(z^2) \quad \text{and} \quad f_2(z) = z f_1(z^2) + f_2(z^2) .$$

Since the only accepting state is state 2, the generating function for the Thue sequence is $f(z) = f_2(z)$. Also

$$f_1(z) + f_2(z) = \frac{1}{1-z} \quad \text{and} \quad f_1(z^2) + f_2(z^2) = \frac{1}{1-z^2} ,$$

and so the previous equation may be written

$$f(z) = z \left(\frac{1}{1-z^2} - f(z^2) \right) + f(z^2) ;$$

this simplifies to

$$f(z) = (1-z)f(z^2) + \frac{z}{1-z^2} \; ,$$

in agreement with the functional equation (6.15).

- For the automaton in figure 6.3 we may write

$$\begin{pmatrix} f_1(z) \\ f_2(z) \\ f_3(z) \end{pmatrix} = \begin{pmatrix} 1 & z & 0 \\ 0 & 0 & 1 \\ z & 1 & z \end{pmatrix} \begin{pmatrix} f_1(z^2) \\ f_2(z^2) \\ f_3(z^2) \end{pmatrix} \; .$$

The generating function of the DFA is $f(z) = f_1(z) + f_3(z)$ and we can, for example, eliminate f_3 to give

$$f_1(z) = f_1(z^2) + zf_2(z^2) \; , \quad f(z) = f_1(z^2) + zf(z^2) + (1+z)f_2(z^2) \; ;$$

since

$$f(z^2) + f_2(z^2) = \frac{1}{1-z^2}$$

we then obtain

$$f_1(z) = f_1(z^2) - zf(z^2) + \frac{z}{1-z^2} \; , \quad f(z) = f_1(z^2) - f(z^2) + \frac{1}{1-z} \; .$$

The price we pay for reducing the number of equations from three to two is that the homogeneous system has turned into a non–homogeneous system; moreover, it appears to be impossible to derive a single functional equation for $f(z)$ in terms of $f(z^2)$. To give transcendence proofs for numbers derived from arbitrary automata we would need to consider matrix functional equations such as (6.19).

- For the DFA on page 166 we might also seek to prove transcendence of the "decimal"

$$1.32331232313311233123132\cdots$$

in any base $r \geq 4$. Here, instead of just writing down a digit 1 in the kth place when the DFA accepts k and a 0 when it does not, we have written down the state arrived at when k is input. This number is $f(1/r)$, where

$$f(z) = f_1(z) + 2f_2(z) + 3f_3(z) \; ,$$

and again we could find a system of functional equations involving $f(z)$ and some of the other $f_m(z)$.

6.6   CONCLUSION

In this chapter we have seen that certain decimals in which the digits form a simple "pattern" can be generated as the output of a deterministic finite automaton. We can use the automaton (or other means) to write the given decimal in the form $f(1/b)$ and find an identity relating $f(z)$ and $f(z^r)$, or a system of such identities; and in certain cases we can use these functional equations to prove the given decimal transcendental.

EXERCISES

6.1 Prove the third part of Lemma 6.2: if $\beta_1, \beta_2, \ldots, \beta_m$ are algebraic numbers, then
$$\|\beta_1 \beta_2 \cdots \beta_m\| \leq \|\beta_1\| \|\beta_2\| \cdots \|\beta_m\| .$$
Give one example where equality does not hold, and one where it does. Also, show that the inequality
$$\|\beta_1 + \beta_2\| \leq \|\beta_1\| + \|\beta_2\|$$
is not always true.

6.2 *An alternative definition of algebraic size.* If α is a root of the irreducible polynomial $f(z) = a_n z^n + a_{n-1} z^{n-1} + \cdots + a_0$ with relatively prime integer coefficients, define
$$\|\alpha\| = 1 + H(f) = 1 + \max_{k=0,1,\ldots,n} |a_k| .$$
Prove that any non–zero algebraic number α satisfies $|\alpha| \, \|\alpha\| \geq 1$, but that the stronger inequality $|\alpha| \, H(f) \geq 1$ is not always true.

6.3 *Proving a function to be transcendental.* Let $p(z)$ be a (complex) polynomial with degree k and leading coefficient p_k. Show that if c_1, c_2 are positive real constants with $c_1 < |p_k| < c_2$, then
$$c_1 |z|^k < |p(z)| < c_2 |z|^k$$
whenever $|z|$ is sufficiently large. Use this to prove that the exponential function, $f(z) = e^z$, is a transcendental function.

6.4 For any non–negative integer k, let $a_k = 1$ if k, written in base 10, contains the digits 0 and 1 only, and $a_k = 0$ otherwise. Prove that the real number α with decimal digits a_k, in any base $b \geq 2$, that is,
$$\alpha = a_0.a_1 a_2 a_3 a_4 \cdots = 1.100000000110000 \cdots = \sum_{k=0}^{\infty} \frac{a_k}{b^k} ,$$
is transcendental.

6.5 Let r be an integer, $r \geq 2$, and write
$$S_r = \{\, k \in \mathbb{N} \mid k \text{ has an odd number of digits in base } r \,\} .$$
Prove that
$$\alpha = \sum_{k \in S_r} \frac{1}{m^k}$$
is transcendental for any integer $m \geq 2$.

6.6 Let

$$f(z) = z + 2z^2 + 2z^3 + 4z^4 + 4z^5 + 4z^6 + 4z^7 + 8z^8 + 8z^9 + \cdots ,$$

where the coefficients consist of a 1, two 2s, four 4s, eight 8s and so on. Prove that if ζ is algebraic and $0 < |\zeta| < 1$, then $f(\zeta)$ is transcendental.

6.7 Consider

$$f(z) = z + 2z^2 + 2z^3 + 3z^4 + 3z^5 + 3z^6 + 3z^7 + 4z^8 + 4z^9 + \cdots ,$$

where the coefficients consist of a 1, two 2s, four 3s, eight 4s and so on. By relating $f(z)$ to a function we have already studied in this chapter, show that $f(\zeta)$ is transcendental for non–zero algebraic ζ inside the unit circle.

6.8 Let

$$f(z) = z^2 - z^3 - z^4 + z^5 + z^6 + z^7$$
$$- z^8 - z^9 - z^{10} - z^{11} - z^{12} - z^{13} + z^{14} + z^{15} + \cdots ;$$

the signs (starting from the z^2 term) consist of a plus, 2 minuses, 3 pluses, 6 minuses, 9 pluses, 18 minuses and so on. Prove that if ζ is algebraic and $0 < |\zeta| < 1$, then $f(\zeta)$ is transcendental.

6.9 Suppose that $f(z)$ is a function which is analytic for $|z| < 1$ and satisfies the functional equation

$$f(z^4) = (1 + z)f(z)f(z^2) .$$

with $f(0) = 1$. Suppose further that $f(z)$ is a transcendental function, and let ζ be a non–zero algebraic number with $|\zeta| < 1$. Prove that at least one of $f(\zeta)$ and $f(\zeta^2)$ is a transcendental number.

6.10 Let S be the set of non–negative integers whose binary representation includes (at least) two consecutive ones. That is,

$$S = \{\, 11,\ 110,\ 111,\ 1011,\ 1100,\ 1101,\ 1110,\ 1111,\ 10011, \dots \,\}$$
$$= \{\, 3,\ 6,\ 7,\ 11,\ 12,\ 13,\ 14,\ 15,\ 19, \dots \,\}$$

in binary and decimal notation respectively.

(a) Construct a DFA which accepts S.

(b) If $f(z)$ is the characteristic function of S, find a linear system of functional equations relating the values of f and other functions at z to their values at z^2. Use as few other functions as you can.

APPENDIX 1: ALPHABETS, LANGUAGES AND DFAS

An **alphabet** is a finite set. If Σ is an alphabet, then a **word** over Σ is any finite sequence of elements of Σ. A word is customarily written with neither brackets nor commas, $w = a_1 a_2 \cdots a_n$ where the symbols a_k are elements of Σ. The **length** of the word just described is n. There is a unique word ε of length zero, called the **empty word**. We denote by Σ^* the set of all words over Σ. A **language** over Σ is a subset of Σ^*.

A **deterministic finite automaton** $\mathcal{M}$ over an alphabet Σ is a quadruple $\langle Q, q_1, \delta, F \rangle$, where

- Q is a finite set consisting of the **states** of $\mathcal{M}$;

- q_1 is an element of Q, the **initial state**;

- δ is a function from $Q \times \Sigma$ to Q, known as the **transition function**;

- $F \subseteq Q$ is the set of **final** or **accepting** states.

The **extended transition function** $\delta^* : Q \times \Sigma^* \to Q$ is defined inductively:

$$\delta^*(q, \varepsilon) = q \quad \text{and} \quad \delta^*(q, wa) = \delta\big(\delta^*(q, w), a\big)$$

for all $q \in Q$, $w \in \Sigma^*$ and $a \in \Sigma$.

A word $w \in \Sigma^*$ is said to be **accepted**, or **recognised** by $\mathcal{M}$ if and only if $\delta^*(q_1, w) \in F$, and the subset of Σ^* given by

$$L(\mathcal{M}) = \{ w \in \Sigma^* \mid w \text{ is accepted by } \mathcal{M} \}$$

is the language accepted by $\mathcal{M}$.

APPENDIX 2: SOME RESULTS OF COMPLEX ANALYSIS

A 2.1   Taylor series and analytic functions

If a function f is analytic in an open disc D with centre z_0, then it has a Taylor series

$$\sum_{k=0}^{\infty} a_k (z - z_0)^k , \tag{6.19}$$

with $a_k = f^{(k)}(z_0)/k!$, which converges to $f(z)$ in D. Conversely, if a series such as (6.19) converges in an open disc D with centre z_0, then it represents a function $f(z)$ analytic in D; moreover, the coefficients a_k are precisely the Taylor series coefficients of f.

Proof. See, for example, Churchill and Brown [19], sections 57, 65 and 66.

A 2.2   Limit points of roots of an analytic function

Suppose that $f(z)$ is analytic in a disc $D \subseteq \mathbb{C}$, and that f has a sequence of roots $w_0, w_1, w_2, \ldots$ which converges to a limit w in D. Then f is identically zero in D.

Proof. The zeros of an analytic function, other than the zero function, are isolated ([19], section 75). But under the stated conditions f is analytic and therefore continuous at w, and we have

$$f(w) = f\left(\lim_{k\to\infty} w_k\right) = \lim_{k\to\infty} f(w_k) = 0 \ .$$

Thus w is a root of f and is not isolated since every neighbourhood of w contains a point w_k which is also a root of f. Hence f is identically zero, as claimed.

A 2.3   Estimation of power series

The absolute value of a (convergent) power series can be estimated from its first non–zero term alone. More precisely, let $\varepsilon > 0$ and suppose that the power series

$$\sum_{k=K}^{\infty} a_k z^k$$

converges in the disc $|z| < R + \varepsilon$. Then there is a constant c such that for all z with $0 < |z| \le R$, we have

$$\left|\sum_{k=K}^{\infty} a_k z^k\right| \le c |z|^K \ . \tag{6.20}$$

Note that the constant c does not depend on z, but may depend on R and on the particular power series we are considering.

Proof. Denote the given series by $f(z)$. Since it converges for $|z| < R + \varepsilon$, so does the series

$$\sum_{k=K}^{\infty} a_k z^{k-K} \ .$$

But as we have seen above, a convergent Taylor series represents an analytic function; moreover, an analytic function on a closed disc $|z| \le R$ is bounded ([19], section 18); so for all z in this disc, we have

$$|f(z)| = |z|^K \left|\sum_{k=K}^{\infty} a_k z^{k-K}\right| \le c |z|^K$$

as claimed.

Example. For the well–known series

$$\sin z = \sum_{k=0}^{\infty} (-1)^k \frac{z^{2k+1}}{(2k+1)!}$$

we have $K = 1$. The series converges for all complex z; choose, for instance, $R = 3$. If $|z| \le R$, then

$$\left| \frac{\sin z}{z} \right| = \left| \sum_{k=0}^{\infty} (-1)^k \frac{z^{2k}}{(2k+1)!} \right| \le \sum_{k=0}^{\infty} \frac{3^{2k}}{(2k+1)!} = \frac{\sinh 3}{3} < 3.3392 \cdots .$$

Thus

$$|\sin z| \le 4 \, |z| \quad \text{whenever} \quad 0 < |z| \le 3 .$$

Note that the familiar real inequality $|\sin z| \le |z|$ is *not* generally true for complex z.

A 2.4   Algebraic and transcendental functions

To show that the function f defined in equation (6.1) on page 149 is transcendental, we may use two of many results on this topic to be found in George Pólya and Gabor Szegő's classic text *Problems and Theorems in Analysis* [51], Part VIII, Chapter 3, section 4.

Lemma 6.7. *Suppose that the coefficients of the power series*

$$f(z) = \sum_{k=0}^{\infty} a_k z^k$$

are integers and have only finitely many different values. Then f is a rational function if and only if the sequence of coefficients is eventually periodic.

Comment. Once again we see a strong analogy between rational *numbers* and rational *functions*.

Lemma 6.8. *Let f have a power series with integral coefficients. If f is an algebraic function but not a rational function, then the series has radius of convergence strictly less than 1.*

Corollary 6.9. *The function f that we considered on pages 149–156 is transcendental.*

Proof. The power series coefficients of f are

$$a_k = \begin{cases} 1 & \text{if } k \text{ is a power of 2} \\ 0 & \text{if not,} \end{cases}$$

so the first lemma implies that f is not a rational function. But the ratio test shows that the series for f has radius of convergence equal to 1, and therefore by the second lemma f cannot be an algebraic function. Hence f is transcendental.

In the present case the non–vanishing of E can be proved by conceptually simpler, though more laborious, methods: we may expand the sum defining $E(z)$, and the increasingly large gaps in the series for $f(z)$ will enable us to explicitly identify a non–zero term in the expansion.

Let p be maximal such that $a_p(z)$ is not the zero polynomial, and let q be maximal such that the coefficient of z^q in $a_p(z)$ is non–zero. Choose r such that $s < 2^r$; the significance of this choice will appear later. Expanding

$$a_p(z)f(z)^p = (a_{p0} + \cdots + a_{pq}z^q)(z + z^2 + z^4 + z^8 + \cdots)^p \qquad (6.21)$$

we obtain, among others, a term

$$p!\, a_{pq}z^q\, z^{2^r} z^{2^{r+1}} z^{2^{r+2}} \cdots z^{2^{r+p-1}}, \qquad (6.22)$$

where the $p!$ is due to the $p!$ different orders in which p different terms z^{2^k} can be chosen from the p repetitions of the second factor in (6.21). The coefficient of this term is not zero. We shall show that

- if we expand the expression (6.21) there are no terms with the same exponent as (6.22), except for those we have already counted; and

- if we expand $a_j(z)f(z)^j$ for $0 \le j < p$ there are no such terms at all.

It will follow that the term (6.22) in the power series of $E(z)$ is not cancelled by any other term, and so $E(z)$ is not identically zero. **Notation.** We shall write S_j for a sum of j powers of 2.

So, firstly, we expand (6.21) and seek terms in z^k, with

$$k = q + 2^r + 2^{r+1} + 2^{r+2} + \cdots + 2^{r+p-1} = q + 2^{r+p} - 2^r .$$

Any such term has an exponent of the form $q' + S_p$ with $0 \le q' \le q$. So we require $q' + S_p = q + 2^{r+p} - 2^r$, that is,

$$S_p = 2^{r+p} - 2^r + q - q' .$$

Since $0 \le q' \le q \le s < 2^r$, this equality implies

$$2^{r+p} - 2^r \le S_p < 2^{r+p} .$$

But if $S_p < 2^{r+p}$, we have

$$S_p \le 2^{r+p-1} + 2^{r+p-2} + \cdots + 2^r = 2^{r+p} - 2^r \le S_p ,$$

and so the only solution is the one we have already. Similarly, if we expand $a_j(z)f(z)^j$ we want

$$q' + S_j = q + 2^{r+p} - 2^r$$

with $0 \le q' \le s$ and $0 \le j \le p - 1$. Now

$$-2^r < -s \le q - q' \le s < 2^r$$

and so

$$2^{r+p} - 2^{r+1} = 2^{r+p} - 2^r - 2^r < S_j < 2^{r+p} ;$$

as above we have

$$S_j \le S_{p-1} \le 2^{r+p-1} + 2^{r+p-2} + \cdots + 2^{r+1} = 2^{r+p} - 2^{r+1} ,$$

and the contradiction between this inequality and the previous one shows that no term can be found with exponent the same as that in (6.22). Therefore, $E(z)$, upon expansion, contains a term (6.22) which is not cancelled out by any other term, and so $E(z)$ does not vanish identically.

Comment. It is comparatively simple to show that f is not a rational function. Suppose that $f(z) = p(z)/q(z)$, where p and q are polynomials of degrees m and n respectively. Then the functional equation (6.2) can be rewritten as the polynomial identity

$$p(z)q(z^2) - p(z^2)q(z) = zq(z)q(z^2) ; \qquad (6.23)$$

the three terms have degrees $m + 2n$, $2m + n$ and $3n + 1$ respectively.

- If $m + 2n \ge 2m + n$, then $n \ge m$ and so $3n + 1 > m + 2n$. Thus the right–hand side has greater degree than the left.
- If $m + 2n < 2m + n$, then $n < m$, so $n + 1 \le m$ and $3n + 1 \le m + 2n$. In this case the left–hand side has greater degree than the right.

Thus no equation such as (6.23) can hold, and f is not a rational function.

For an **alternative proof**, once again suppose $f(z) = p(z)/q(z)$; without loss of generality we may assume that the polynomials p and q have no common roots. From the identity (6.23), we have for any $\alpha \in \mathbb{C}$ that

$$q(\alpha) = 0 \quad \text{if and only if} \quad q(\alpha^2) = 0 . \qquad (6.24)$$

It follows that any root α of $q(z)$ must satisfy $\alpha = 0$ or $|\alpha| = 1$, as otherwise $q(z)$ would have infinitely many roots $\alpha, \alpha^2, \alpha^4, \ldots$. If $\alpha = 0$, then 0 is, say, an n–fold root of the left–hand side in (6.23) and a $(3n + 1)$–fold root of the right–hand side: this is impossible. If $q(z)$ has roots on the unit circle, choose a root $e^{i\theta}$ with minimal positive θ; then (6.24) shows that $e^{i\theta/2}$ is also a root, contradicting minimality. Thus $q(z)$ cannot have any roots at all, and must be a constant polynomial; but then the left–hand side and right–hand side of (6.23) have different degrees. We have a contradiction, and so f is not a rational function.

APPENDIX 3: A RESULT ON LINEAR EQUATIONS

If $m < n$, then a homogeneous system of m linear equations in n unknowns, with coefficients in the rational field $\mathbb{Q}$, always has a non–zero solution in $\mathbb{Q}^n$.

Proof. Any system of linear equations over $\mathbb{Q}$ has no solution, a unique solution or infinitely many solutions in $\mathbb{Q}^n$. In the present case the system, being homogeneous, cannot fail to have solutions; and since the number of variables exceeds the number of equations, the general solution must contain one or more parameters. So the system has infinitely many solutions, and hence at least one non–zero solution.

CHAPTER 7

Lambert's Irrationality Proofs

> *Proving that the ratio of the diameter of a circle to its*
> *circumference is not rational will not surprise geometers*
> *...but what merits more attention,*
> *and will be rather a greater surprise, is that*
> *if the ratio of an arc of a circle to its radius is rational,*
> *then the ratio of the tangent to the radius is not.*
>
> J.H. Lambert [37]

IN THE 1760s J.H. Lambert proved the irrationality of π, e and related numbers by means of continued fractions. Roughly, the methods used are those we used in Chapter 4 to find the continued fraction of e; however, more general continued fractions must be employed. For numbers related to e and to π we shall need expressions of the forms

$$a_1 + \cfrac{b_1}{a_2 + \cfrac{b_2}{a_3 + \cdots}} \qquad \text{and} \qquad a_1 - \cfrac{b_1}{a_2 - \cfrac{b_2}{a_3 - \cdots}}$$

respectively, where in each case the successive numerators and denominators will be positive integers. In the first case the convergence of the (infinite) continued fraction, and the behaviour of its convergents, follow patterns quite similar to those for simple continued fractions, and all the properties we shall need are proved with little additional difficulty. The second case is significantly harder. The main problem is to ensure that the continued fraction converges – not only in the infinite case, but also in the finite case, for it is by no means clear that attempting to calculate an expression such as

$$\cfrac{b_1}{a_1 - \cfrac{b_2}{a_2 - \cdots - \cfrac{b_n}{a_n}}}$$

DOI: 10.1201/9781003111207-7

will not at some stage lead to a zero denominator. In fact, finding and proving general conditions which ensure that such expressions are meaningful turns out to be so messy that we shall use the continued fractions only as an inspiration for defining certain sequences which will correspond to convergents.

We begin by using the functions $f(c; z)$ from Chapter 4 to develop continued fractions for $\tan z$ and related functions. First, recall that

$$(\tfrac{1}{2})^{(k)} = \frac{(2k)!}{2^{2k}k!} \quad \text{and} \quad (\tfrac{3}{2})^{(k)} = (2k+1)(\tfrac{1}{2})^{(k)} = \frac{(2k+1)!}{2^{2k}k!} \ .$$

Using these expressions as on page 97, we have

$$f(\tfrac{1}{2}; \tfrac{1}{4}z^2) = \cosh z \quad \text{and} \quad zf(\tfrac{3}{2}; \tfrac{1}{4}z^2) = \sinh z \ ;$$

similarly,

$$f(\tfrac{1}{2}; -\tfrac{1}{4}z^2) = \sum_{k=0}^{\infty} \frac{(-1)^k z^{2k}}{(\tfrac{1}{2})^{(k)} 2^{2k}k!} = \sum_{k=0}^{\infty} \frac{(-1)^k z^{2k}}{(2k)!} = \cos z$$

and

$$zf(\tfrac{3}{2}; -\tfrac{1}{4}z^2) = z\sum_{k=0}^{\infty} \frac{(-1)^k z^{2k}}{(\tfrac{3}{2})^{(k)} 2^{2k}k!} = \sum_{k=0}^{\infty} \frac{(-1)^k z^{2k+1}}{(2k+1)!} = \sin z \ .$$

The identity

$$f(c; z) = f(c+1; z) + \frac{z}{c(c+1)} f(c+2; z)$$

from page 95 can be rearranged as

$$\frac{f(c+1; z)}{2cf(c; z)} = \cfrac{1}{2c + \cfrac{4z}{2c+2} \cfrac{f(c+2; z)}{f(c+1; z)}} \ ; \qquad (7.1)$$

iterating this expression (and ignoring, for the time being, any convergence problems), gives the continued fraction

$$\frac{f(c+1; z)}{2cf(c; z)} = \frac{1}{2c+} \ \frac{4z}{2c+2+} \ \frac{4z}{2c+4+} \ \frac{4z}{2c+6+} \ \cdots \ .$$

Now taking $c = \tfrac{1}{2}$, replacing z by $\tfrac{1}{4}z^2$ and multiplying both sides by z yields

$$\tanh z = \frac{z}{1+} \ \frac{z^2}{3+} \ \frac{z^2}{5+} \ \frac{z^2}{7+} \ \cdots \ ,$$

which agrees with Chrystal [18], Part II, Chapter XXXIV, section 21, equation (16). We can employ a similar procedure to find the relevant continued fraction when z is written explicitly as a rational number s/t. Divide both sides of (7.1) by t to give

$$\frac{1}{2ct} \frac{f(c+1; z)}{f(c; z)} = \cfrac{1}{2ct + \cfrac{4t^2 z}{(2c+2)t} \cfrac{f(c+2; z)}{f(c+1; z)}} \ ;$$

iterate and multiply by s to yield

$$\frac{s}{2ct}\frac{f(c+1;z)}{f(c;z)}$$

$$= \frac{s}{2ct +}\frac{4t^2z}{(2c+2)t +}\frac{4t^2z}{(2c+4)t +}\cdots\frac{4t^2z}{(2c+2k-2)t + \sigma_k}, \qquad (7.2)$$

where

$$\sigma_k = \frac{4t^2z}{(2c+2k)t}\frac{f(c+k+1;z)}{f(c+k;z)}.$$

Finally, take $c = \frac{1}{2}$ and $z = s^2/4t^2$ to obtain

$$\tanh\frac{s}{t} = \frac{s}{t +}\frac{s^2}{3t +}\frac{s^2}{5t +}\frac{s^2}{7t +}\cdots. \qquad (7.3)$$

We repeat that to do this properly we have to prove that the right–hand side of (7.3) does actually converge, and, moreover, that it converges to the value claimed. By very similar calculations (just replace z by $-\frac{1}{4}z^2$ instead of $\frac{1}{4}z^2$), and subject to the same cautions, we have

$$\tan z = \frac{z}{1 -}\frac{z^2}{3 -}\frac{z^2}{5 -}\frac{z^2}{7 -}\cdots \quad \text{and} \quad \tan\frac{s}{t} = \frac{s}{t -}\frac{s^2}{3t -}\frac{s^2}{5t -}\frac{s^2}{7t -}\cdots.$$

7.1 GENERALISED CONTINUED FRACTIONS

Now we need to work out the theory of continued fractions having the form of the first expression on page 177; we must pay particular attention to the question of convergence of infinite continued fractions. We shall assume that all a_k and b_k are positive real numbers (later we shall assume further that they are in fact integers); it is then clear that any *finite* expression

$$\frac{b_1}{a_1 +}\frac{b_2}{a_2 +}\cdots\frac{b_k}{a_k}$$

is meaningful, and we define

$$\frac{b_1}{a_1 +}\frac{b_2}{a_2 +}\frac{b_3}{a_3 +}\cdots = \lim_{k\to\infty}\frac{b_1}{a_1 +}\frac{b_2}{a_2 +}\cdots\frac{b_k}{a_k},$$

provided that the limit exists. The first few truncations of the finite continued fraction are

$$\frac{b_1}{a_1}, \quad \frac{b_1}{a_1 +}\frac{b_2}{a_2} = \frac{a_2b_1}{a_2a_1 + b_2}, \quad \frac{b_1}{a_1 +}\frac{b_2}{a_2 +}\frac{b_3}{a_3} = \frac{a_3a_2b_1 + b_3b_1}{a_3a_2a_1 + a_3b_2 + b_3a_1}$$

and so on, from which it is fairly easy to guess that we should define

$$p_k = a_kp_{k-1} + b_kp_{k-2} \quad \text{and} \quad q_k = a_kq_{k-1} + b_kq_{k-2}$$

with the initial conditions $p_{-1} = q_0 = 1$ and $p_0 = q_{-1} = 0$. Note that if every b_k is 1, then these formulae coincide with those for the convergents of a simple continued fraction with $a_0 = 0$.

Now let x, like a_k and b_k, be a positive real number. By methods essentially identical to those of Chapter 4 we may show that

$$\cfrac{b_1}{a_1 + } \ \cdots \ \cfrac{b_{k-1}}{a_{k-1} + } \ \cfrac{b_k}{x} = \cfrac{b_1}{a_1 + } \ \cdots \ \cfrac{b_{k-1}}{a_{k-1} + b_k/x} = \frac{xp_{k-1} + b_k p_{k-2}}{xq_{k-1} + b_k q_{k-2}}$$

and hence that

$$\cfrac{b_1}{a_1 + } \ \cfrac{b_2}{a_2 + } \ \cdots \ \cfrac{b_k}{a_k} = \frac{p_k}{q_k} . \tag{7.4}$$

A closely related result which will be of use in the future is

$$\cfrac{b_1}{a_1 + } \ \cdots \ \cfrac{b_{k-1}}{a_{k-1} + } \ \cfrac{b_k}{a_k + x} = \frac{(a_k + x)p_{k-1} + b_k p_{k-2}}{(a_k + x)q_{k-1} + b_k q_{k-2}} = \frac{p_k + xp_{k-1}}{q_k + xq_{k-1}} .$$

It is also easy to prove by induction that

$$p_{k-1}q_k - p_k q_{k-1} = (-1)^k b_k b_{k-1} \cdots b_1 . \tag{7.5}$$

Since a_k and b_k are positive, the convergent

$$\frac{p_k}{q_k} = \frac{a_k p_{k-1} + b_k p_{k-2}}{a_k q_{k-1} + b_k q_{k-2}}$$

lies strictly between p_{k-1}/q_{k-1} and p_{k-2}/q_{k-2}; it is easy to check that the second convergent is less than the first, and so

$$\frac{p_2}{q_2} < \frac{p_4}{q_4} < \frac{p_6}{q_6} < \cdots < \frac{p_5}{q_5} < \frac{p_3}{q_3} < \frac{p_1}{q_1} . \tag{7.6}$$

Now suppose that a_k and b_k are positive integers, and that $a_k \geq b_k$ for sufficiently large k, say for all $k \geq K$. It is then clear that whenever $k \geq 3$, we have $q_{k-1} > q_{k-2}$ and so $q_k > (a_k + b_k)q_{k-2}$; moreover,

$$q_k > (2b_k)q_{k-2} > (2b_k)(2b_{k-2})q_{k-4} > \cdots > (2b_k)(2b_{k-2}) \cdots ,$$

where the last factor in the product is either $2b_{K+2}q_K$ or $2b_{K+1}q_{K-1}$. Hence for $k \geq K$, we have

$$\left| \frac{p_k}{q_k} - \frac{p_{k-1}}{q_{k-1}} \right| = \frac{b_k b_{k-1} \cdots b_1}{q_k q_{k-1}} < \frac{b_k b_{k-1} \cdots b_1}{2^{k-K} b_k b_{k-1} \cdots b_{K+1} q_K q_{K-1}} = \frac{2^K b_K \cdots b_1}{2^k q_K q_{K-1}} ;$$

since K is a fixed number, the right–hand side tends to zero as $k \to \infty$. Combining this with (7.6) and (7.4) proves the first part of the following result.

Theorem 7.1. Irrationality of a generalised continued fraction. *If $a_1, a_2, \ldots$ and $b_1, b_2, \ldots$ are positive integers with $a_k \geq b_k$ for all sufficiently large k, then the infinite continued fraction*

$$\alpha = \frac{b_1}{a_1 +} \; \frac{b_2}{a_2 +} \; \frac{b_3}{a_3 +} \; \cdots \tag{7.7}$$

converges to an irrational limit.

Proof. We have shown that the required limit exists; it remains to prove this limit irrational. So, suppose that α is rational. For $k \geq 1$ write

$$\alpha_k = \frac{b_k}{a_k +} \; \frac{b_{k+1}}{a_{k+1} +} \; \frac{b_{k+2}}{a_{k+2} +} \; \cdots \; ,$$

and note that if α_k is a rational number p/q, then

$$\alpha_{k+1} = \frac{b_k}{\alpha_k} - a_k = \frac{b_k q - a_k p}{p} \tag{7.8}$$

is also rational. Since $\alpha_1 = \alpha$ is rational, so is every α_k. But for sufficiently large k we have $a_k \geq b_k$, and the inequalities (7.6) yield

$$\alpha_k < \frac{b_k}{a_k} \leq 1 \; ;$$

so in (7.8), we have $p < q$, and the denominators of the α_k eventually form an infinite decreasing sequence of positive integers. This is impossible, and so α cannot be rational.

Comment. The condition that $a_k \geq b_k$ for sufficiently large k is by no means necessary in order that the continued fraction converge. For example, Chrystal [18], Part II, Chapter XXXIV, section 14 shows that the continued fraction (7.7) converges if the series

$$\sum_{k=2}^{\infty} \frac{a_{k-1} a_k}{b_k}$$

diverges. It follows from standard series convergence tests that the continued fraction converges if

$$\frac{a_{k-1} a_k}{b_k} \nrightarrow 0 \quad \text{as } k \to \infty$$

or if

$$\lim_{k \to \infty} \frac{a_{k+1} b_k}{a_{k-1} b_{k+1}} > 1 \; .$$

Chrystal also gives a necessary and sufficient convergence criterion: the continued fraction (7.7) is convergent if and only if at least one of the series

$$\frac{a_1}{b_1} + \frac{a_3 b_2}{b_3 b_1} + \frac{a_5 b_4 b_2}{b_5 b_3 b_1} + \frac{a_7 b_6 b_4 b_2}{b_7 b_5 b_3 b_1} + \cdots \quad \text{and} \quad \frac{a_2 b_1}{b_2} + \frac{a_4 b_3 b_1}{b_4 b_2} + \frac{a_6 b_5 b_3 b_1}{b_6 b_4 b_2} + \cdots$$

is divergent. Observe that (7.6) holds whenever all a_k and b_k are positive. Therefore, it is impossible for a continued fraction to be unboundedly divergent; it will fail to converge only if

$$\lim_{k\to\infty} \frac{p_{2k}}{q_{2k}} < \lim_{k\to\infty} \frac{p_{2k+1}}{q_{2k+1}}.$$

In this case the continued fraction is said to **oscillate**. By Chrystal's criterion above, a simple example of an oscillating continued fraction is

$$\frac{2}{1+}\ \frac{4}{1+}\ \frac{8}{1+}\ \frac{16}{1+}\ \cdots.$$

It does not seem easy to evaluate the two limit points of this continued fraction. However, there is a very simple result for a closely related expression: for the continued fraction

$$\frac{4}{1+}\ \frac{6}{1+}\ \frac{10}{1+}\ \frac{18}{1+}\ \cdots\ \frac{2^k+2}{1+}\ \cdots$$

we have

$$\lim_{k\to\infty} \frac{p_{2k}}{q_{2k}} = 1 \quad \text{and} \quad \lim_{k\to\infty} \frac{p_{2k+1}}{q_{2k+1}} = 2,$$

as shown in David Angell and Michael D. Hirschhorn [7].

7.1.1 Irrationality of $\tanh r$

To prove the irrationality of $\tanh r$ for non–zero rational r we must consider the question of convergence in the derivation of the continued fraction (7.3). Taking, again, $c = \frac{1}{2}$ and $z = s^2/4t^2$ in (7.2), we have for each k an identity

$$\tanh\frac{s}{t} = \frac{s}{t+}\ \frac{s^2}{3t+}\ \frac{s^2}{5t+}\ \cdots\ \frac{s^2}{(2k-1)t+\sigma_k},$$

where

$$\sigma_k = \frac{s^2}{(2k+1)t}\ \frac{f(k+\frac{3}{2};s^2/4t^2)}{f(k+\frac{1}{2};s^2/4t^2)}.$$

It is clear from the defining series that $f(c;z)$ is positive whenever c and z are positive; therefore every σ_k is positive, and the following theorem may be applied.

Theorem 7.2. *Let $\alpha \in \mathbb{R}$, let $a_1, a_2, \ldots, b_1, b_2, \ldots \in \mathbb{R}^+$, and suppose that there exist positive real numbers $\sigma_1, \sigma_2, \ldots$ such that*

$$\alpha = \frac{b_1}{a_1+}\ \frac{b_2}{a_2+}\ \cdots\ \frac{b_k}{a_k+\sigma_k} \tag{7.9}$$

for each $k \geq 1$. If the infinite continued fraction

$$\frac{b_1}{a_1+}\ \frac{b_2}{a_2+}\ \frac{b_3}{a_3+}\ \cdots$$

converges, then its limit is α.

Proof. Suppose that the continued fraction converges to β; that is, the convergents p_k/q_k tend to β as $k \to \infty$. Then from

$$\alpha = \cfrac{b_1}{a_1 +} \cfrac{b_2}{a_2 +} \cdots \cfrac{b_k}{a_k + \sigma_k} = \frac{p_k + \sigma_k p_{k-1}}{q_k + \sigma_k q_{k-1}}$$

we see by using the result in appendix 4.1 (repeated at the end of the present chapter) that for every k the constant α lies between p_k/q_k and p_{k-1}/q_{k-1}. By the Sandwich Theorem $\alpha = \beta$, which is what we wanted to prove.

Corollary 7.3. *If r is rational and not zero, then $\tanh r$ is irrational.*

Proof. Let $r = s/t$ be a non–zero rational; since the hyperbolic tangent is an odd function, we may assume that $s > 0$. Clearly $(2k-1)t > s^2$ for sufficiently large k, so by Theorem 7.1, the continued fraction

$$\cfrac{s}{t +} \cfrac{s^2}{3t +} \cfrac{s^2}{5t +} \cfrac{s^2}{7t +} \cdots$$

converges to an irrational limit. But by the theorem just proved and the remarks preceding it this limit is $\tanh r$; the result follows.

Corollary 7.4. *If r is a non–zero rational, then e^r is irrational. In particular, e is irrational.*

Proof. Use the previous corollary and the identity

$$\tanh \frac{r}{2} = \frac{e^r - 1}{e^r + 1}.$$

Corollary 7.5. *If r is a positive rational with $r \neq 1$, then $\log r$ is irrational (where, as usual, $\log$ denotes the natural logarithm).*

Proof. Use the relation

$$\tanh(\log r) = \frac{r^2 - 1}{r^2 + 1}.$$

Comment. Theorem 7.2 makes it appear that any choice of partial numerators and denominators is possible in the generalised continued fraction of a given real number. This is not quite true, as a bad choice will force $\sigma_k < 0$ for some k, which is not permissible. Nonetheless a wide variety of generalised continued fractions exists converging to a given limit. If, however, the numerators b_k are given, then the a_k are more or less uniquely determined, as the following result shows.

Proposition 7.6. *If α is irrational, if the b_k are given integers, and if we demand that the a_k are integers such that for each k a relation of the form (7.9) holds with $0 < \sigma_k < 1$, then the a_k are uniquely determined.*

Proof. We have $\alpha = b_1/(a_1 + \sigma_1)$ with $0 < \sigma_1 < 1$; so

$$a_1 < a_1 + \sigma_1 = \frac{b_1}{\alpha} < a_1 + 1$$

and a_1 is the integer part, σ_1 the fractional part, of b_1/α. Similarly, a_{k+1} is the integer part of b_{k+1}/σ_k for each $k \geq 1$.

Example. With the assistance of the computer algebra system `Maple` we find

$$\sqrt{2} = \cfrac{2}{1 +} \cfrac{1}{2 +} \cfrac{2}{4 +} \cfrac{1}{1 +} \cfrac{2}{9 +} \cfrac{1}{1 +} \cfrac{2}{3 +} \cfrac{1}{1 +} \cfrac{2}{9 +} \cfrac{1}{1 +} \cfrac{2}{3 +} \cdots$$

$$= \cfrac{2}{1 +} \cfrac{3}{7 +} \cfrac{4}{16 +} \cfrac{5}{10 +} \cfrac{6}{19 +} \cfrac{7}{8 +} \cfrac{8}{8 +} \cfrac{9}{20 +} \cfrac{10}{31 +} \cfrac{11}{86 +} \cdots$$

$$= \cfrac{5}{3 +} \cfrac{3}{5 +} \cfrac{5}{8 +} \cfrac{8}{26 +} \cfrac{26}{647 +} \cfrac{647}{712 +} \cfrac{712}{1019 +} \cfrac{1019}{3233 +} \cfrac{3233}{21848 +} \cdots$$

$$= \cfrac{2}{1 +} \cfrac{4}{9 +} \cfrac{8}{12 +} \cfrac{16}{89 +} \cfrac{32}{125 +} \cfrac{64}{111 +} \cfrac{128}{142 +} \cfrac{256}{758 +} \cfrac{512}{541 +} \cdots .$$

7.2 FURTHER CONTINUED FRACTIONS

To prove the irrationality of $\tan r$ for non–zero rational r we study generalised continued fractions having the form of the second expression on page 177. As remarked earlier, it is difficult to ensure rigorously that such expressions converge, even in the finite case, so we shall work directly with an appropriate sequence of fractions p_k/q_k.

Given two infinite sequences $a_1, a_2, \ldots$ and $b_1, b_2, \ldots$ of positive real numbers, we define

$$p_k = a_k p_{k-1} - b_k p_{k-2} \quad \text{and} \quad q_k = a_k q_{k-1} - b_k q_{k-2}$$

with initial conditions $p_0 = q_{-1} = 0$, $p_{-1} = -1$ and $q_0 = 1$. Informally

$$\frac{p_k}{q_k} = \frac{b_1}{a_1 -} \frac{b_2}{a_2 -} \cdots \frac{b_k}{a_k}, \tag{7.10}$$

though we shall not actually seek to prove anything like this. Our aim is to show that

- under suitable conditions on a_k and b_k, the quotient p_k/q_k tends to an irrational limit as $k \to \infty$;

- certain specific values of a_k and b_k satisfy these conditions;

- for these specific values, the limiting value of p_k/q_k is $\tan r$, where r is a given non–zero rational number.

We begin by making the following assumption, which will remain in force until it is superseded by a weaker assumption in Theorem 7.17.

Assumption. For all k, the denominators a_k and numerators b_k are positive integers with $a_k > b_k + 1$.

Comment. In fact, the assumption $a_k > b_k$ suffices to prove many properties of the continued fractions (7.10); however, the above assumption is sufficient for our purposes and simplifies our arguments.

First, we ensure that, subject to the above assumption, the denominator q_k never vanishes.

Lemma 7.7. *For each $k \geq 0$, we have $q_k > q_{k-1} \geq 0$.*

Proof. The case $k = 0$ is immediate from the definition; proceeding inductively we have

$$q_k = a_k q_{k-1} - b_k q_{k-2} \geq (b_k + 1)q_{k-1} - b_k q_{k-2}$$
$$= q_{k-1} + b_k(q_{k-1} - q_{k-2}) > q_{k-1} \geq 0$$

for $k \geq 1$.

Corollary 7.8. *If $k \geq 0$, then q_k is strictly positive; moreover, $q_k \to \infty$ as $k \to \infty$.*

The following lemma is easily proved.

Lemma 7.9. *If a_k and b_k are positive real numbers, and p_k and q_k are defined as above, then*

$$p_k q_{k-1} - p_{k-1} q_k = b_k b_{k-1} \cdots b_1$$

for each $k \geq 1$.

Corollary 7.10. *The fraction p_k/q_k increases monotonically with k.*

Lemma 7.11. *Let $r_k = q_k - p_k$. Then $r_k \geq r_{k-1}$ for each $k \geq 0$.*

Proof by induction. The basis of the induction is clear. If we take $k \geq 1$ and assume that the result is true for $1, 2, \ldots, k - 1$, then we have the inequalities $r_{k-1} \geq r_{k-2} \geq \cdots \geq r_{-1} = 1$; hence

$$r_k = a_k r_{k-1} - b_k r_{k-2} \geq (b_k + 1)r_{k-1} - b_k r_{k-1} = r_{k-1} \ ,$$

and the induction proceeds.

Corollary 7.12. *If $k \geq 0$, then $p_k/q_k < 1$.*

Corollary 7.13. *As $k \to \infty$, the fraction p_k/q_k tends to a limit α with $0 < \alpha \leq 1$.*

Next we wish to show that the limit α is irrational. In effect, we shall do this by writing

$$\alpha_m = \cfrac{b_m}{a_m -} \ \cfrac{b_{m+1}}{a_{m+1} -} \ \cdots = \frac{b_m}{a_m - \alpha_{m+1}}$$

and mimicking the proof of Theorem 7.1. However, in order to avoid any convergence problems we once again refrain from employing this sort of continued fraction, and instead work directly with p_k/q_k and related fractions. We denote by $p_{m,k}$ and $q_{m,k}$ the numbers satisfying the same initial conditions and recurrences as p_k and q_k, except that the partial numerators $b_1, b_2, \dots$ and denominators $a_1, a_2, \dots$ are replaced by $b_m, b_{m+1}, \dots$ and $a_m, a_{m+1}, \dots$ respectively. That is,

$$q_{m,-1} = 0 \ , \quad q_{m,0} = 1 \ , \quad q_{m,1} = a_m \ , \quad q_{m,2} = a_{m+1}a_m - b_{m+1}$$

and so on. Specifically, the recurrence for $p_{m,k}$ is

$$p_{m,k} = a_{m+k-1}p_{m,k-1} - b_{m+k-1}p_{m,k-2} \ ,$$

and that for $q_{m,k}$ is similar. Clearly $p_{1,k} = p_k$ and $q_{1,k} = q_k$ for all k. Since the partial numerators and denominators defining $p_{m,k}$ and $q_{m,k}$ are positive integers satisfying $a_k > b_k + 1$, the properties proved above for p_k and q_k also hold for $p_{m,k}$ and $q_{m,k}$. In particular, the quotient $p_{m,k}/q_{m,k}$ tends to a limit α_m as $k \to \infty$, and this limit satisfies $0 < \alpha_m \leq 1$.

The proof of the following lemma is an easy induction on k.

Lemma 7.14. *For all integers $m \geq 1$ and $k \geq -1$, we have*

$$p_{m,k+1} = b_m q_{m+1,k} \quad and \quad q_{m,k+1} = a_m q_{m+1,k} - p_{m+1,k} \ .$$

Corollary 7.15. *For each $m \geq 1$, we have*

$$\alpha_m = \frac{b_m}{a_m - \alpha_{m+1}} \ .$$

Proof. By the arguments given for p_k and q_k, all $p_{m,k}$ and $q_{m,k}$ are positive for $k \geq 1$. From the lemma, we have

$$\frac{p_{m,k+1}}{q_{m,k+1}}\left(a_m - \frac{p_{m+1,k}}{q_{m+1,k}}\right) = \frac{b_m q_{m+1,k}}{a_m q_{m+1,k} - p_{m+1,k}} \cdot \frac{a_m q_{m+1,k} - p_{m+1,k}}{q_{m+1,k}} = b_m \ ;$$

taking the limit as $k \to \infty$ of each side yields $\alpha_m(a_m - \alpha_{m+1}) = b_m$, and since $b_m \neq 0$ the desired equality follows.

Theorem 7.16. *The limit α of p_k/q_k is irrational.*

Proof. First, observe that since $\alpha_{m+1} \leq 1$, we have

$$\alpha_m = \frac{b_m}{a_m - \alpha_{m+1}} \leq \frac{b_m}{a_m - 1} \ ,$$

and so in fact each α_m is *strictly* less than 1. For any m, if α_m is a rational number p/q, then

$$\alpha_{m+1} = a_m - \frac{b_m}{\alpha_m} = \frac{pa_m - qb_m}{p} \ ,$$

which has smaller denominator than α_m. If $\alpha = \alpha_1$ is rational, therefore, the denominators of the α_m constitute an infinite decreasing sequence of positive integers. This is impossible, and so α must be irrational.

We now have to prove the existence and irrationality of α without assuming that $a_k > b_k + 1$ for all k.

Theorem 7.17. *Suppose that a_k and b_k are positive integers for all $k \geq 1$, and that $a_k > b_k + 1$ for all sufficiently large k. Then*

$$\alpha = \lim_{k \to \infty} \frac{p_k}{q_k}$$

exists and is an irrational number.

Proof. Suppose that the inequality relating a_k and b_k holds for all $k \geq K$. Since the numerators $b_K, b_{K+1}, \ldots$ and denominators $a_K, a_{K+1}, \ldots$ all satisfy this inequality, our previous results show that $p_{K,k}/q_{K,k}$ tends to an irrational limit α_K as $k \to \infty$. By induction on k we can show that

$$\begin{aligned}
p_{m,k} &= -p_{m,\ell}\, p_{m+\ell+1,k-\ell-1} + p_{m,\ell+1}\, q_{m+\ell+1,k-\ell-1}\, , \\
q_{m,k} &= -q_{m,\ell}\, p_{m+\ell+1,k-\ell-1} + q_{m,\ell+1}\, q_{m+\ell+1,k-\ell-1}
\end{aligned} \tag{7.11}$$

for $m \geq 1$ and $0 \leq \ell \leq k$; the induction is facilitated by the observation that the relations are trivial for $\ell = k$ and for $\ell = k - 1$. Taking $m = 1$ and $\ell = K - 2$, we have

$$p_k = -p_{K-2}p_{K,k-K+1} + p_{K-1}q_{K,k-K+1}$$

for $k \geq K - 1$; since $q_{K,k-K+1}$ is not zero,

$$\frac{p_k}{q_{K,k-K+1}} = -p_{K-2}\frac{p_{K,k-K+1}}{q_{K,k-K+1}} + p_{K-1}\, .$$

Treating the second equation of (7.11) in the same way, and then letting k increase without bound, we find that

$$\frac{p_k}{q_{K,k-K+1}} \to p_{K-1} - \alpha_K p_{K-2} \quad \text{and} \quad \frac{q_k}{q_{K,k-K+1}} \to q_{K-1} - \alpha_K q_{K-2}$$

as $k \to \infty$.

Now $q_{K-1} - \alpha_K q_{K-2}$ is not zero. If it were, then the irrationality of α_K would give $q_{K-1} = q_{K-2} = 0$, hence $q_{K-3} = 0$, and eventually $q_0 = 0$, which is false. Consequently q_k is non–zero for sufficiently large k, for if q_k were zero for infinitely many k then $q_k/q_{K,k-K+1}$ would have a limit of zero, or no limit at all. Therefore, for sufficiently large k we have

$$\frac{p_k}{q_k} = \frac{p_k/q_{K,k-K+1}}{q_k/q_{K,k-K+1}} \to \frac{p_{K-1} - \alpha_K p_{K-2}}{q_{K-1} - \alpha_K q_{K-2}}\, ,$$

and this is the desired limit α. To show that α is irrational, suppose otherwise; then the relation

$$\alpha_K(p_{K-2} - \alpha q_{K-2}) = p_{K-1} - \alpha q_{K-1} \,,$$

with the irrationality of α_K, implies that $p_{K-1} - \alpha q_{K-1} = 0 = p_{K-2} - \alpha q_{K-2}$. But this implies

$$\begin{aligned}
0 &= (p_{K-1} - \alpha q_{K-1})q_{K-2} - (p_{K-2} - \alpha q_{K-2})q_{K-1} \\
&= p_{K-1}q_{K-2} - p_{K-2}q_{K-1} \\
&= b_{K-1}b_{K-2}\cdots b_1 \,,
\end{aligned}$$

which is impossible since each b_k is non–zero. We conclude that α is irrational, and the proof is complete.

7.2.1 Irrationality of $\tan r$

Now let $r = s/t$ be a non–zero rational number; since the tangent function is odd, we may assume that s and t are positive integers. From our informal derivation of the continued fraction for $\tan r$ on page 179 it is apparent that we should define partial numerators and denominators by

$$a_k = (2k-1)t \quad \text{for } k \geq 1 \quad \text{and} \quad b_k = \begin{cases} s & \text{for } k = 1 \\ s^2 & \text{for } k \geq 2. \end{cases}$$

Consider the three aims stated on page 184. It is clear that $a_k > b_k + 1$ for sufficiently large k, and we have shown that in this case p_k/q_k tends to an irrational limit as $k \to \infty$. It remains to prove that this limit is $\tan r$. Rephrasing the definition of p_k and q_k, we have

$$\begin{aligned}
p_0 &= 0\,, \quad p_1 = s\,, \quad p_{k+1} = (2k+1)tp_k - s^2 p_{k-1}\,, \\
q_0 &= 1\,, \quad q_1 = t\,, \quad q_{k+1} = (2k+1)tq_k - s^2 q_{k-1}
\end{aligned}$$

for $k \geq 1$.

In order to show that $\tan r$ is the limit of p_k/q_k as $k \to \infty$, we write the difference as a fraction,

$$\tan r - \frac{p_k}{q_k} = \frac{q_k \sin r - p_k \cos r}{q_k \cos r}\,,$$

and begin by looking at the numerator.

Lemma 7.18. *If k is a non–negative integer, then*

$$q_k \sin r - p_k \cos r = \sum_{m=0}^{\infty} (-1)^m \frac{2^k (m+k)!}{m!\,(2m+2k+1)!} \frac{s^{2m+2k+1}}{t^{2m+k+1}}\,.$$

Moreover, $q_k \sin r - p_k \cos r \to 0$ as $k \to \infty$.

Proof. The series formula is proved by induction on k. It is not too hard to check that the result is true for $k = 0$ and for $k = 1$; suppose it is true for some integer $k \geq 1$, and also for $k - 1$. Then we have

$$q_{k+1} \sin r - p_{k+1} \cos r = (2k + 1)t(q_k \sin r - p_k \cos r)$$
$$- s^2(q_{k-1} \sin r - p_{k-1} \cos r)$$
$$= \sum_{m=0}^{\infty} (-1)^m (2k + 1) \frac{2^k (m + k)!}{m! (2m + 2k + 1)!} \frac{s^{2m+2k+1}}{t^{2m+k}}$$
$$- \sum_{m=0}^{\infty} (-1)^m \frac{2^{k-1} (m + k - 1)!}{m! (2m + 2k - 1)!} \frac{s^{2m+2k+1}}{t^{2m+k}} .$$

Writing this expression as a single sum yields

$$\sum_{m=0}^{\infty} (-1)^m \frac{2^{k-1} (m + k)!}{m! (2m + 2k + 1)!} \left[2(2k+1) - \frac{(2m + 2k + 1)(2m + 2k)}{m + k} \right] \frac{s^{2m+2k+1}}{t^{2m+k}} ,$$

and upon simplification the factor in square brackets is just $-4m$. Therefore, we may drop the $m = 0$ term and then shift the summation index to give

$$q_{k+1} \sin r - p_{k+1} \cos r = \sum_{m=1}^{\infty} (-1)^{m+1} \frac{2^{k+1} (m + k)!}{(m - 1)! (2m + 2k + 1)!} \frac{s^{2m+2k+1}}{t^{2m+k}}$$
$$= \sum_{m=0}^{\infty} (-1)^m \frac{2^{k+1} (m + k + 1)!}{m! (2m + 2k + 3)!} \frac{s^{2m+2k+3}}{t^{2m+k+2}} .$$

By induction, the identity is true for all $k \geq 0$. Now the ratio of successive terms in the series has absolute value

$$\frac{(m + k + 1)!}{(m + 1)! (2m + 2k + 3)!} \frac{s^{2m+2k+3}}{t^{2m+k+3}} \bigg/ \frac{(m + k)!}{m! (2m + 2k + 1)!} \frac{s^{2m+2k+1}}{t^{2m+k+1}}$$
$$= \frac{1}{2(m + 1)(2m + 2k + 3)} \frac{s^2}{t^2} < \frac{s^2}{4kt^2} .$$

Thus for large k (specifically, for $k > s^2/4t^2$), we have an alternating series in which the terms, right from the very first, are monotonically decreasing in absolute value. Therefore

$$0 < q_k \sin r - p_k \cos r < \{ \text{first term} \} = \frac{2^k k!}{(2k + 1)!} \frac{s^{2k+1}}{t^{k+1}} < \frac{s}{t} \frac{(s^2/t)^k}{k!} .$$

The right–hand side tends to zero as $k \to \infty$, and so $q_k \sin r - p_k \cos r \to 0$, as claimed.

We also need to know what happens to q_k for large k. We showed in Corollary 7.2 that $q_k \to \infty$ as $k \to \infty$; but this was proved only under the assumption that $a_k > b_k + 1$ for all k, which need not be the case here. Indeed, if we take $s = 2$ and $t = 1$ it is easy to calculate $q_2 = -1$ and $q_3 = -9$; for $k \geq 3$ we have $(2k + 1)t > s^2 > 0$, so

$$q_{k+1} = (2k + 1)tq_k - s^2 q_{k-1} < q_k$$

and $q_k \to -\infty$ as $k \to \infty$.

Lemma 7.19. *If $a_k > b_k + 1$ for all sufficiently large k, then either $q_k \to \infty$ or $q_k \to -\infty$ as $k \to \infty$.*

Proof. With the notation $q_{m,k}$ introduced on page 186, we can use (7.11) as in the proof of Theorem 7.17 to show that

$$q_k = q_{K,k-K+1}\left(q_{K-1} - \frac{p_{K,k-K+1}}{q_{K,k-K+1}} q_{K-2}\right)$$

for $k \geq K - 1$. Continuing with the ideas of this proof, we see that as $k \to \infty$ the bracketed factor tends to a non–zero limit; moreover, $q_{K,k-K+1} \to \infty$ since $a_k > b_k + 1$ for all $k \geq K$. Therefore, q_k tends to ∞ or $-\infty$, depending on whether the bracketed quantity tends to a positive or negative limit.

We can now complete the irrationality proof for $\tan r$. Assume that r is not an odd multiple of $\frac{1}{2}\pi$. (Of course we "know" that this is true since π is irrational – but part of our current aim is to give an alternative proof of that same fact!) Then $\cos r \neq 0$, and by the two lemmas we have just proved,

$$\tan r - \frac{p_k}{q_k} = \frac{q_k \sin r - p_k \cos r}{q_k \cos r} \to 0 \qquad (7.12)$$

as $k \to \infty$; that is,

$$\lim_{k \to \infty} \frac{p_k}{q_k} = \tan r .$$

Corollary 7.20. *If r is rational, not zero, and not an odd multiple of $\frac{1}{2}\pi$, then $\tan r$ is irrational.*

Comment. We mention parenthetically that (7.12) enables us very easily to settle the question of whether q_k tends to ∞ or to $-\infty$. The expression $\tan r - p_k/q_k$ is always decreasing (Corollary 7.10) and tends to zero, so it is always positive; and we know that $q_k \sin r - p_k \cos r$ is positive; therefore q_k tends to ∞ if $\cos r$ is positive, and to $-\infty$ if $\cos r$ is negative.

Finally we observe that we can now prove π to be irrational, and as a consequence can remove the extraneous condition in the above corollary.

Corollary 7.21. π *is irrational.*

Proof. Suppose that π is rational. Then $\frac{1}{4}\pi$ is rational; it certainly is neither zero nor an odd integer times $\frac{1}{2}\pi$, and so by the previous corollary $1 = \tan \frac{1}{4}\pi$ is irrational. But this is false.

Theorem 7.22. *If r is a non–zero rational number, then $\tan r$ is irrational.*

EXERCISES

7.1 Given positive integers $a_1, a_2, \ldots$ and $b_1, b_2, \ldots$, define p_k and q_k by the recurrences and initial conditions given in section 7.1:

$$p_k = a_k p_{k-1} + b_k p_{k-2} \quad \text{and} \quad q_k = a_k q_{k-1} + b_k q_{k-2}$$

with $p_{-1} = q_0 = 1$ and $q_{-1} = p_0 = 0$. Show how p_k and q_k may be evaluated in terms of a matrix product similar to that in exercise 4.5.

7.2 Prove that

$$\frac{2}{1+} \frac{3}{2+} \frac{4}{3+} \frac{5}{4+} \frac{6}{5+} \cdots = 1$$

by showing that its convergents p_k/q_k satisfy

$$q_k = (k+1)! - k! + (k-1)! - (k-2)! + \cdots \pm 1!$$

and $p_k = q_k - (-1)^k$.

7.3 Show that

$$\frac{1}{2+} \frac{2}{2+} \frac{3}{3+} \frac{4}{4+} \cdots = \frac{1}{e},$$

a result originally due to Euler.

7.4 Prove that

$$\log 2 = \frac{1}{1+} \frac{1}{1+} \frac{4}{1+} \frac{9}{1+} \cdots$$

where log denotes the natural logarithm; and that

$$\tan^{-1} x = \frac{x}{1+} \frac{x^2}{3 - x^2 +} \frac{9x^2}{5 - 3x^2 +} \frac{25x^2}{7 - 5x^2 +} \cdots$$

for $|x| \le 1$. Use the latter to obtain a continued fraction for π.

7.5 Define p_k and q_k as in section 7.2,

$$p_k = a_k p_{k-1} - b_k p_{k-2} \quad \text{and} \quad q_k = a_k q_{k-1} - b_k q_{k-2}$$

with $p_{-1} = q_0 = 1$ and $q_{-1} = p_0 = 0$; assume that a_k and b_k are positive integers with $a_k > b_k + 1$ for all k.

(a) Prove that if $k \ge 2$, then $q_k > (b_k + b_{k-1})q_{k-2}$.

(b) Show that the inequality $q_k > 2b_k q_{k-2}$ is *not* generally true.

7.6 Prove that if r^2 is rational and not zero, then $r \tan r$ is irrational. Deduce that π^2 is irrational.

7.7 For $k \geq 1$, let

$$a_k = \frac{8k^2 + 1}{2k} \left(\frac{(2k-3)(2k-7)\cdots}{(2k-1)(2k-5)\cdots} \right)^2 ,$$

where the dots indicate the product of a decreasing arithmetic progression with difference 4, continuing as long as the factors are positive. An empty product is taken to be 1, so that we have $a_1 = \frac{9}{2}$. Show that

$$2 - \frac{1}{a_1 -} \; \frac{1}{a_2 -} \; \frac{1}{a_3 -} \; \cdots = \frac{\pi}{2} .$$

APPENDIX: SOME RESULTS FROM ELEMENTARY ALGEBRA AND CALCULUS

A simple property of positive fractions. If a, b, c and d are positive and a/b is not equal to c/d, then

$$\frac{a+c}{b+d}$$

lies (strictly) between a/b and c/d.

The following result is known variously as the Sandwich Theorem, the Squeeze Theorem, or the Pinching Theorem.

Theorem 7.23. *Let* $\{u_k\}$, $\{v_k\}$ *and* $\{w_k\}$ *be sequences of real numbers. If*

$$u_k \leq v_k \leq w_k \text{ for all } k \quad \text{and} \quad \lim_{k \to \infty} u_k = \lim_{k \to \infty} w_k = L ,$$

then the limit as $k \to \infty$ *of* v_k *exists and is equal to* L.

Corollary 7.24. *If* $u_k \to L$ *as* $k \to \infty$, *and if for each* k *the constant* α *lies between* u_{k-1} *and* u_k, *then* $L = \alpha$.

> *May not music be described as the mathematics of the sense,*
> *mathematics as music of the reason?*
> *The musician feels mathematics, the mathematician thinks music:*
> *music the dream, mathematics the working life.*
>
> J.J. Sylvester

s for exercises

hat $(2q - p)/(p - q) = p/q$ and note that $0 < p - q < q$.

ing that β_k is an integer, show that $q' = q\beta_{k+1} - q\lfloor\beta_{k+1}\rfloor$ and $q'\alpha$
ch integers and that $0 \le q' < q$.

nsider $(a - b)/(\sqrt{a} - \sqrt{b})$.

p/q, where p, q are coprime, then $px + qy = 1$ for some integers
se this to show that $(p/q)^{1/p}$ is rational and deduce that $p = m^p$
he positive integer m. When is this possible?

y assume that a, b, c are integers having no common factor. Now
lying by $\sqrt[3]{2}$ we get $2c + a\sqrt[3]{2} + b\sqrt[3]{4} = 0$, and then another similar
on. This gives a homogeneous system of three linear equations in
ariables having a non–trivial solution; so the determinant of the
must be zero. Working it out,

$$a^3 + 2b^3 + 4c^3 - 6abc = 0 ,$$

lowing the argument for the irrationality of $\sqrt{2}$ yields a contra-
. There is a much easier solution using methods from Chapter 3.

vith the fact that $q^2 - 2p^2$ is a non–zero integer, so $|q^2 - 2p^2| \ge 1$.
roblem looks forward to topics we shall discuss in Chapters 3

rational, then $(3 - b)x$ and $(2 - a)/x$ are also rational.

ormula for $\sin 3\theta$ to find a polynomial $p(z)$ with integer coefficients
α as a root. Don't do any more work than you have to in ruling
ssible rational values of α! Better still, do something similar for
ead of α.

a well–known relation connecting $\cos^2 r\pi$ and $\cos 2r\pi$.

1.14 The solution to exercise 1.13 relies implicitly on the fact that there is a power of 2 with any specified number of digits. This is not true for powers of 13, so a more subtle argument will be required.

1.16 Write the (eventually periodic) decimal of α as

$$\alpha = 0.d_1 d_2 \cdots d_m e_1 \cdots e_p e_1 \cdots e_p \cdots .$$

Explain why if $n > m$, then a_{n+p} has more digits than a_n, and deduce that $a_{n+2p} > 10 a_n$. Choose

$$x = 10^{1/2p} \quad \text{and} \quad 0 < c < \min\left\{ \frac{a_k}{x^k} \,\middle|\, k = 1, 2, \ldots, m + 2p \right\} .$$

1.18 (a) Denote the sum of the digits of a positive integer k by $\sigma(k)$. Then, most of the time, $\sigma(k+1)$ is equal to $\sigma(k) + 1$: when is it not?

(b) Suppose that $\alpha = 0.d_1 d_2 \cdots$ is rational and that its decimal has period p. Choose a multiple q of p such that $\sigma(q)$ is not a multiple of 10, and choose a such that $10^t \equiv a \pmod{p}$ for infinitely many t. For any sufficiently large such t, we have

$$\sigma(10^t) \equiv 1 \pmod{10} , \quad \sigma(10^t + q) \not\equiv 1 \pmod{10} ,$$

which means that the sequence $\{ d_{mp+a} \}_{m \geq 0}$ contains each of two different values infinitely often. But this is impossible.

1.19 Using the Maclaurin series for $\cos x$ gives $(2n)! \cos 1 = N \pm R$, where N is an integer and

$$R = \frac{1}{(2n+1)(2n+2)} - \cdots .$$

1.20 Suppose that $b = ae + ce^{-1}$ with $a, b, c \in \mathbb{Z}$; we may assume that $a > 0$ and $c \neq 0$. If $c > 0$ and n is odd, we have

$$bn! - a(n! + \cdots + 1) - c(n! - \cdots - 1) = ar_n + cs_n$$

with

$$r_n = \frac{n!}{(n+1)!} + \frac{n!}{(n+2)!} + \cdots , \quad s_n = \frac{n!}{(n+1)!} - \frac{n!}{(n+2)!} + \cdots ;$$

this gives $0 < ar_n + cs_n < (a+c)/n$.

1.21 Suppose that $\alpha = p/q$. Explain why there exists n such that $g_1 g_2 \cdots g_n$ is a multiple of q; then write $g_1 g_2 \cdots g_n \alpha$ as an integer plus a remainder, and use the standard type of argument.

1.22 Use the exponential series to write $A = (2n)!\, \alpha/2^{n+1}$ as an integer plus a remainder. Then find n such that A is also an integer. This idea is due to D.W. Masser.

1.23 All dihedral angles of a regular tetrahedron, and of a cube, are the same: $\cos^{-1} \frac{1}{3}$ for the tetrahedron, $\frac{\pi}{2}$ for the cube. Problem 1.12 tells us something useful about rational values of $\cos x$.

1.24 Suppose that $x > y > 0$. If $x^2 \in \mathbb{Q}$ and $y^2 \in \mathbb{Q}$ and $x/y \notin \mathbb{Q}$, then $(x - y)^2 \notin \mathbb{Q}$.

Euclid also has the assumption $x^2/y^2 \in \mathbb{Q}$, but this clearly follows from the first two hypotheses.

CHAPTER 2

2.1 Use integration by parts twice to show that

$$
I_{n+2} = \frac{(4n + 6)(n + 2)}{\pi^2} I_{n+1} - \frac{4(n + 2)(n + 1)}{\pi^2} I_n \,,
$$

with initial conditions $I_0 = 0$ and $I_1 = 4/\pi^2$. Supposing that $\pi^2 = a/b$, define $J_n = a^n I_n/n!$; prove that J_n is always an integer, that J_n is non–zero for infinitely many n, and that J_n tends to zero as $n \to \infty$.

2.2 We need to prove that if $\pi = a/b$, then

$$
\left| \int_0^{\pi/2} (a^2 - b^2 x^2)^n \cos x \, dx \right| > \left| \int_{\pi/2}^{\pi} (a^2 - b^2 x^2)^n \cos x \, dx \right| .
$$

Show that this is a consequence of the fact that $0 < x < \pi - x$ whenever $0 < x < \frac{\pi}{2}$.

2.3 Suppose that $\pi\sqrt{c} = a/b$ and take

$$
f(x) = x^n (a - b\sqrt{c}\, x)^n \,.
$$

As in the proof of Theorem 2.4 we obtain

$$
I = F(\pi) + F(0) \quad \text{with} \quad F(x) = f(x) - f''(x) + \cdots \,.
$$

Noting that $f(x) = f(\pi - x)$ gives the useful simplification $I = 2F(0)$. Taking a bit of care (since f does not have integral coefficients) we can show that if n, k are even and $k > n$, then $f^{(k)}(0)$ is an integer multiple of $(n + 1)!$; this leads in the usual way to a contradiction for sufficiently large even n. The irrationality of π^2 follows on noting that if $\pi^2 = p/q$, then $\pi\sqrt{pq} = p$.

2.4 Apply the ideas in the proof of Theorem 2.5 to $\displaystyle\int_0^r f(x) \sinh x \, dx$.

2.5 First, prove that

$$I = \frac{1}{r} F(0) - \frac{1}{r} F(1) \cos r$$

where

$$F(x) = f(x) - \frac{1}{r^2} f''(x) + \frac{1}{r^4} f^{(4)}(x) - \cdots + \frac{1}{r^{4n}} f^{(4n)}(x) .$$

Supposing $r^2 = a/b$, show that $a^{2n} F(1)$ is an integer divisible by $(n+1)!$ and that $a^{2n} F(0)$ is a non–zero multiple of $n!$.

2.6 (a) Induction. The initial conditions are $u_0 = v_1 = 0$, $u_1 = v_0 = 1$.

(b) First, prove that $b^n d^n u_n$ and $b^n d^n v_n$ are integers.

(c) Use the integral formula to estimate $|J_{\nu+n+t}(r)|$. The estimate involves a gamma function term, which can be related to a factorial.

(d) Don't forget to show that infinitely many $J_{\nu+n}(r)$ are non–zero.

(e) Simplifying

$$\Gamma(k + \tfrac{1}{2}) = (k - \tfrac{1}{2})(k - \tfrac{3}{2}) \cdots (\tfrac{3}{2})(\tfrac{1}{2})\Gamma(\tfrac{1}{2})$$

shows that the series for $J_{1/2}(x)$ and $J_{-1/2}(x)$ are closely related to well–known Maclaurin expansions.

2.7 (a) The given series converges for all z, so f is an entire function and has a Taylor series centred at any $z_0 \in \mathbb{C}$. The integral formula is a standard result of Taylor series.

(b) Use the given series to confirm the differential equation; substitute the other series into it to find

$$4(n + 2)(n + 1)c_{n+2} + 2(2n + 1)(n + 1)c_{n+1} - r^2 c_n = 0$$

for $n \geq 0$, with

$$c_0 = f(1) = \cosh r \quad \text{and} \quad c_1 = f'(1) = \frac{r \sinh r}{2} .$$

(c) Find a recurrence for d_n. If even two consecutive d_n are zero, then using the recurrence "backwards" shows that all d_n are zero, which is not the case.

(d) The given series shows immediately that $|f(z)| \leq \cosh(r|z|^{1/2})$.

(e) If z is on the circle of integration, then $|z| \leq 1+n^2 \leq 4n^2$; estimating the integral gives $|c_n| \leq e^{2rn}/n^{2n}$.

CHAPTER 3

3.1 The minimal polynomial is

$$f(z) = z^6 - 6z^4 - 6z^3 + 12z^2 - 36z + 1 .$$

To prove that this polynomial is irreducible we can reduce modulo 3 to get $f_3(z) = z^6 + 1 = (z^2 + 1)^3$; this shows that $f(z)$ has no linear or cubic factors. It remains to eliminate the possibility that $f(z)$ is the product of a quadratic and a quartic.

3.2 Use the triple–angle formula $\cos 3\theta = 4\cos^3\theta - 3\cos\theta$ to show that α satisfies $f(z) = 8z^3 - 6z - 1$. The value of $\cos 3\theta$ is unchanged if θ is replaced by $\theta + \frac{2}{3}k\pi$; so the conjugates of α are $\cos\frac{\pi}{9}$, $\cos\frac{7\pi}{9}$ and $\cos\frac{13\pi}{9}$.

3.3 Remember that $1 + \zeta + \zeta^2 + \zeta^3 + \zeta^4 = 0$ and $\zeta^5 = 1$. We have

$$\alpha = \zeta + \zeta^4 , \quad \alpha^2 = \zeta^2 + 2 + \zeta^3 \quad \Rightarrow \quad \alpha^2 + \alpha = 1 ,$$

so α is a root of $f(z) = z^2 + z - 1$. Doing something similar for β gives the polynomial $g(z) = z^4 + 2z^3 + 4z^2 + 3z + 1$; consider g modulo 2 to show that it is irreducible.

3.4 If $f(z) = a_n z^n + a_{n-1}z^{n-1} + \cdots + a_1 z + a_0$, then

$$a_0 z^n + a_1 z^{n-1} + \cdots + a_{n-1}z + a_n = z^n f\left(\frac{1}{z}\right) .$$

3.5 (a) Use De Moivre's Theorem to find a formula for $\sin n\theta$, then divide by $\cos^n\theta$ and take $\theta = \pi/n$ to show that $\tan(\pi/n)$ is a root of

$$f_n(z) = \sum_{\substack{j=1 \\ j \text{ odd}}}^{n} (-1)^{(j-1)/2}\binom{n}{j} z^{j-1} .$$

(b) If $n = p$ is prime, then the binomial coefficient $\binom{p}{j}$ is a multiple of p for $j = 1, 2, \ldots, p-1$. Hence

$$f_p(z) = z^{p-1} - \binom{p}{p-2}z^{p-3} + \cdots \pm \binom{p}{3}z^2 \mp \binom{p}{1}$$

is irreducible by Eisenstein's criterion.

(c) If n is composite and has an odd prime factor p, then $\tan(\pi/p)$ is a root of $f_n(z)$, so $f_p(z)$ is a factor of $f_n(z)$, so $f_n(z)$ is not irreducible. The polynomial $f_{10}(z)$ has degree 8 and it has $f_5(z)$ as a factor; the quotient $5z^4 - 10z^2 + 1$ is the minimal polynomial of $\tan(\pi/10)$. It can be proved irreducible by using exercise 3.4.

3.6 Let $S = \{d \in \mathbb{Z} \mid d\alpha$ is an algebraic integer $\}$. If d_1, d_2 are in S, then $d_1 - d_2$ is in S; it follows that S is the set of all multiples of some integer d, and this d is the denominator of α.

3.7 The $\cos 3\theta$ formula shows that α is a root of $f(z) = 20z^3 - 15z - 4$. To show efficiently that $f(z)$ is irreducible, use an idea from an example on page 4. Find the conjugates of α as in exercise 3.2; alternatively, divide $f(z)$ by $z - \alpha$ and solve a quadratic. By problem 3.6, den α is a factor of 20; confirm that $d\alpha$ is an algebraic integer for $d = 10$ and not for any smaller factor.

3.8 Write $\alpha = a^{1/n}$ and $c = a^{1/n} - b^{1/n}$; show that the polynomials

$$f(z) = z^n - a \quad \text{and} \quad g(z) = (z - c)^n - b$$

have rational coefficients, that α is a root of each, and that they have no other common roots. Deduce that α is rational.

3.9 If 1, $\sqrt[3]{2}$, $\sqrt[3]{4}$ are linearly dependent, then $\sqrt[3]{2}$ is a root of a rational quadratic.

3.10 (a) Suppose that α and $\alpha + r$ are conjugates; let $f(z)$ be their minimal polynomial. Since α is a root of $f(z + r)$ we have $f(z) = f(z + r)$, and comparing coefficients gives $r = 0$.

3.11 Suppose that β is a repeated root of $f(z)$. Then β is a root of the derivative $f'(z)$, which is a non–zero rational polynomial; so the minimal polynomial $g(z)$ of β has degree less than that of $f(z)$ and is a factor of $f(z)$. Since $f(z)$ is irreducible, this is impossible.

3.12 Suppose that

$$f(z) = g(z)h(z) = (g_m z^m + \cdots + g_0)(h_m z^m + \cdots + h_0) \,.$$

By following the proof of Eisenstein's Lemma we have (without loss of generality) that $g_0, g_1, \ldots, g_{n-2}$ are all multiples of p. But the leading coefficient of $g(z)$ is ± 1, so $g(z)$ has degree $n - 1$ and $h(z)$ has degree 1. Thus $f(z)$ has a rational root. But this is not so.

3.13 (a) We have $z^4 + 1 = (z^2 + az + 1)(z^2 - az + 1)$ modulo p.

(b) In this case $z^4 + 1 = (z^2 + az - 1)(z^2 - az - 1)$ modulo p...

(c) ... and here $z^4 + 1 = (z^2 + a)(z^2 - a)$ modulo p.

(d) Try the first composite number you think of!

(e) Consider $f(z + 1)$.

3.15 If $f(z) = r(z)s(z)$, then for each k we have $r(a_k)s(a_k) = -1$ and therefore $r(a_k) = -s(a_k)$; hence $r(z) + s(z)$ is a non–zero polynomial with n distinct roots, which must have degree at least n; so the assumed factorisation of f is trivial.

3.16 Consider the polynomial $(z - \alpha)(z - \beta) = z^2 - (\alpha + \beta)z + \alpha\beta$.

3.17 If α is not approximable to order s, then for every c, say for simplicity $c = 1$, the inequalities have only finitely many solutions. Now consider c', the minimum over all these solutions of $q^s|\alpha - p/q|$.

3.18 If

$$0 < \left|\alpha - \frac{p}{q}\right| < \frac{c}{q^s}$$

then

$$\left|\alpha + \frac{p}{q}\right| \leq |2\alpha| + \left|\alpha - \frac{p}{q}\right| \leq |2\alpha| + c$$

and so

$$\left|\alpha^2 - \frac{p^2}{q^2}\right| \leq \frac{c(|2\alpha| + c)}{(q^2)^{s/2}} .$$

3.19 For any m, we have

$$\alpha = \sum_{k=1}^{m} \frac{1}{a^{s^k}} + R = \frac{p}{q} + R$$

with $q = a^{s^m}$ and $0 < R < 2/q^s$.

3.20 Use a "two–dimensional" version of the ideas in the proof of Lemma 3.20. **Comment.** By actually going through the calculations, $k = 41$ is the smallest multiplier that will work.

3.21 For any real $s > 0$ there exist infinitely many m such that $b_{m+1}/b_m > s$, and for any such m we can show

$$0 < \left|\alpha - \frac{p}{q}\right| < \frac{2}{q^s} \quad \text{with} \quad q = a^{b_m} .$$

3.22 Without loss of generality we consider a real number between 0 and 1, say $x = 0.ddd\cdots$ (the ds need not be all the same). This can be expressed as a sum $x = \alpha + \beta$ with

$$\alpha = 0.d \cdots d0 \cdots 0d \cdots d0 \cdots \quad \text{and} \quad \beta = 0.0 \cdots 0d \cdots d0 \cdots 0d \cdots ,$$

the runs of zeros having lengths approximately 1!, 2!, 3!, 4! and so on. Then, essentially, Liouville's basic proof applies to both α and β.

3.23 (a) To construct a cube twice the volume of a given cube is equivalent to constructing a line segment $\alpha = \sqrt[3]{2}$ times the length of a given segment. This is impossible since the degree of $\sqrt[3]{2}$ is not a power of 2.

(b) Given two lines separated by angle θ, dropping a perpendicular from one to the other, which can be done by ruler and compasses, creates two line segments with lengths in the ratio $\cos\theta$. When $\cos\theta$ is rational, $f(z)$ is a rational polynomial having $\alpha = \cos\frac{1}{3}\theta$ as a root; and if $f(z)$ is irreducible, then α has degree 3. To find a non–trisectable angle θ using what we have done in this question, we need $\cos\theta$ to be rational: if we try to keep things simple by choosing $\theta = r\pi$ with r rational, there are not many options!

For the third celebrated unsolved construction problem of antiquity, see Corollary 5.8.

3.24 Two points $\{P_1, P_2\}$ will not suffice (consider points on the perpendicular bisector of P_1P_2). Choose $P_1 = (0,0)$, $P_2 = (\alpha, 0)$ and $P_3 = (0, \alpha)$, where α is to be chosen later. If there exists (x, y) which is not an irrational distance from any of these points, then

$$x^2 + y^2 = r^2 , \quad (x - \alpha)^2 + y^2 = s^2 , \quad x^2 + (y - \alpha)^2 = t^2$$

for some rational numbers r, s, t; this is true only if

$$4r^2\alpha^2 = (\alpha^2 + r^2 - s^2)^2 + (\alpha^2 + r^2 - t^2)^2 .$$

But we can choose α in such a way that this cannot be true for any rationals r, s, t, and this will ensure that a "bad" point (x, y) cannot exist.

3.25 Assuming that x is not zero, we have

$$0 < \left| \alpha - \frac{y}{x} \right| \left| \alpha^2 + \alpha\frac{y}{x} + \frac{y^2}{x^2} \right| = \left| \frac{c}{bx^3} \right| ,$$

the inequality holding since c is not zero, and

$$\alpha^2 + \alpha\frac{y}{x} + \frac{y^2}{x^2} \geq \frac{3\alpha^2}{4} .$$

Therefore

$$0 < \left| \alpha - \frac{y}{x} \right| < \frac{c'}{|x|^3} ;$$

but α is an algebraic number of degree at most 3, and so in view of Roth's Theorem (or Thue's) these inequalities cannot hold for infinitely many x.

3.26 Work in the complex plane. Show that if the sides of the p–gon, in anticlockwise order, are $a_0, a_1, \ldots, a_{p-1}$, then

$$a_0 + a_1\zeta + a_2\zeta^2 + \cdots + a_{p-1}\zeta^{p-1} = 0 ,$$

where $\zeta = e^{2\pi i/p}$. But ζ is an algebraic number with minimal polynomial $1 + z + z^2 + \cdots + z^{p-1}$, and hence $a_0 = a_1 = a_2 = \cdots = a_{p-1}$.

3.27 For any integer $k \geq 1$, let n_k be the total number of digits in the integers $1, 2, 3, \ldots, 10^k - 1$. By following the procedure in the example, show that

$$|q\xi - p| < \frac{c}{q^t}$$

with $q = (10^k - 1)^2 10^{n_k - 1}$ and

$$t = \frac{9k10^{k-1} - 2k}{(k-1)10^{k-1} + 2k},$$

or something similar, depending on the exact details of your estimates. Then show that $t > 9$ for all sufficiently large k and $t \to 9$ as $k \to \infty$, so that ξ is approximable to order 10 (but not to any higher order: at least, not by using this method).

Liouville's Theorem shows that ξ is not algebraic of degree less than 10, Siegel shows ξ is not algebraic of degree less than 25, and Roth shows that ξ is transcendental.

3.28 Let $s > 0$; for any $m > 2s$, define

$$p_m = (2^{1!} + 1)(2^{2!} + 1) \cdots (2^{m!} + 1)$$

$$q_m = 2^{1! + 2! + \cdots + m!}$$

$$\alpha_m = \frac{p_m}{q_m} = \prod_{k=1}^{m} \left(1 + \frac{1}{2^{k!}}\right),$$

and then use the suggested inequality to show that

$$0 < \left|\alpha - \frac{p_m}{q_m}\right| < \frac{2\alpha}{2^{(m+1)!}} < \frac{2\alpha}{q_m^s}.$$

To prove the inequality, we can use a method suggested by Lovro Soldo: show by induction that all 2^t terms obtained by expanding

$$\prod_{k=m+1}^{m+t} \left(1 + \frac{1}{2^{k!}}\right)$$

are different, so that

$$\prod_{k=m+1}^{m+t} \left(1 + \frac{1}{2^{k!}}\right) = 1 + \frac{1}{2^{(m+1)!}} + \cdots$$

$$< 1 + \frac{1}{2^{(m+1)!}} \left(1 + \frac{1}{2} + \frac{1}{2^2} + \cdots\right).$$

CHAPTER 4

4.1 (a) $[3, 1, 4, 1, 5, 9]$; (b) $[a, 1, 1, 2a, 1, 1, 2a, \ldots]$; (c) $\dfrac{7 + \sqrt{17}}{8}$.

4.2 (a) Use the Euclidean algorithm as usual, noting that the recurrence for p_k gives the quotients and remainders, $p_k = a_k p_{k-1} + p_{k-2}$ and so on: you should get

$$\frac{p_n}{p_{n-1}} = a_n + \frac{1}{a_{n-1} +} \quad \cdots \quad \frac{1}{a_1 +} \frac{1}{a_0}.$$

(b) For the "if" proof, use the uniqueness result, Lemma 4.3.

4.3 Prove by induction that $p_{k+m}q_k - p_k q_{k+m} = (-1)^k Q_m$, where

$$\frac{P_m}{Q_m} = a_{k+1} + \frac{1}{a_{k+2} +} \quad \cdots \quad \frac{1}{a_{k+m}}.$$

4.4 Show that p_n and q_n defined in this way satisfy the same recurrences and initial conditions as p_n and q_n defined in the usual way.

4.5 By induction, the product is $\begin{pmatrix} p_k & p_{k-1} \\ q_k & q_{k-1} \end{pmatrix}$; now take determinants.

4.6 Use exercise 4.2 or exercise 4.5 to evaluate the continued fraction as

$$\frac{p_k^2 + p_{k-1}^2}{p_k q_k + p_{k-1} q_{k-1}}.$$

This fraction is in lowest terms since any prime common factor of the numerator and denominator is also a factor of

$$(p_k^2 + p_{k-1}^2)q_k - p_k(p_k q_k + p_{k-1} q_{k-1}) = \pm p_{k-1}$$

and hence also of p_k.

4.7 Evaluate α from the relation

$$\alpha = a_0 + \frac{1}{a_1 +} \quad \cdots \quad \frac{1}{a_{n-1} +} \frac{1}{a_0 + \alpha},$$

and use problem 4.2(b).

4.8 (a) If α has convergents p_k/q_k, the minimal polynomial is

$$f(z) = q_{n-1}z^2 + (q_{n-2} - p_{n-1})z - p_{n-2} ;$$

evaluate $f(-1)$ and $f(0)$.

(b) Induction, using $\alpha_{k+1}^* = 1/(\alpha_k^* - a_k)$.

(d) Note that

$$\alpha_{m_0} = a_{m_0} + \cfrac{1}{a_{m_0+1} +} \ \cdots \ \cfrac{1}{a_{m_0+n-1} +} \ \cfrac{1}{a_{m_0} + \ \cdots} = \alpha_{m_0+n}$$

and use (c). Starting with the fact that $a_{m+n} = a_m$ is true for $m \geq m_0$, we have shown that it is true for $m \geq m_0 - 1$; repeating the argument m_0 times shows that it is true for $m \geq 0$, and this is what we wanted.

4.9 Use Theorem 4.13.

4.10 (a) Find a formula for q_n/q_{n-1} as in exercise 4.2; hence show that

$$\alpha = \cfrac{1}{a_1 +} \ \cdots \ \cfrac{1}{a_{n-1} +} \ \cfrac{1}{a_n +} \ \cfrac{1}{x} \quad \text{with} \quad x = \frac{q_n - q_{n-1}}{q_n} .$$

(b) If $\beta_m = \sum_{k=0}^{m} \frac{1}{g^{2^k}}$ is the mth partial sum of the series, we can write

$$\beta_m = \frac{p_m}{q_m} , \quad q_m = g^{2^m} , \quad \gcd(p_m, q_m) = 1 .$$

We can then use (a) to find the continued fraction of β_{m+1} from that of β_m. Repeating the procedure indefinitely gives the partial quotients of β, though we must note that the last partial quotient at each step (and only the last) will change at the succeeding step. This yields

$$\beta = \cfrac{1}{g-1 +} \ \cfrac{1}{g+2 +} \ \cfrac{1}{g +} \ \cfrac{1}{g +} \ \cfrac{1}{g-2 +} \ \cfrac{1}{g +} \ \cfrac{1}{g+2 +} \ \cdots$$
$$= \cfrac{1}{b_1 +} \ \cfrac{1}{b_2 +} \ \cdots \ ,$$

where $b_1 = g - 1$, $b_2 = g + 2$, $b_3 = g$; and $b_{2^k} = g$, $b_{2^k+1} = g - 2$ for $k \geq 2$; and

$$b_{2^{k+1}+1-m} = b_m \quad \text{for} \quad k \geq 2 \text{ and } m = 2, 3, \ldots, 2^k - 1 .$$

Applying this recursion gives $b_{2345} = b_{1752} = \cdots = b_9 = g - 2$.

(c) We know that β is irrational and hence is approximable to order 2 or more. Now look carefully at how we proved Theorem 4.16. **Comment.** We shall prove in Chapter 6 that β is transcendental.

(d) We can do something similar for $g = 2$, but we need a version of (a) which applies when $a_n = 1$. Specifically, if $a_n = 1$, then

$$\frac{p_n}{q_n} + \frac{(-1)^n}{q_n^2} = [0, a_1, \ldots, a_{n-2}, a_{n-1}, a_{n-1} + 2, a_{n-2}, \ldots, a_1] .$$

4.11 Prove that the partial denominators of α satisfy $q_k < 10^{2^{k!}}$.

4.12 (a) First, write

$$\left|q\alpha - p - \beta\right| \le \left|q\left(\alpha - \frac{m}{n}\right)\right| + \left|\frac{mq - np - n\beta}{n}\right| .$$

Choose m/n to be an appropriate convergent to α. Then use the Bézout property, Lemma 1.11, to show that there exist integers p, q such that $n \le q < 2n$ and $|mq - np - n\beta| < 1$. This will give

$$\left|q\alpha - p - \beta\right| < \frac{3}{n} ;$$

now go back to the choice of m/n and work out what "appropriate" should mean. Explain carefully why we shall obtain infinitely many possibilities for (p, q).

(b) Given β_1, β_2, use Kronecker's Theorem with suitably chosen values of β and ε to prove the stated result. Do the converse too.

(c) Rephrasing the question, we wish to show that for any positive integer n there exist positive integers p, q such that

$$n\,10^p < 2^q < (n+1)10^p ,$$

that is,

$$\log n < q \log 2 - p < \log(n+1) ,$$

where log denotes the logarithm to base 10. This follows quite easily from (b), though a little care is needed to ensure that p is positive.

4.13 Calculate $\pi^4 = [\,97, 2, 2, 3, 1, 16539, \dots\,]$.

4.14 Suppose that the inequality fails for both p_{k-1}/q_{k-1} and p_k/q_k, and note that

$$\left|\left(\frac{p_k}{q_k} - \alpha\right) + \left(\alpha - \frac{p_{k-1}}{q_{k-1}}\right)\right| = \left|\alpha - \frac{p_k}{q_k}\right| + \left|\alpha - \frac{p_{k-1}}{q_{k-1}}\right| ,$$

equality holding since the two bracketed expressions have the same sign.

4.15 We may calculate

$$\frac{\pi^{12}}{\zeta(12)} = 924041 + \frac{1}{1+}\ \frac{1}{3+}\ \frac{1}{1+}\ \frac{1}{2+}\ \frac{1}{2+}\ \frac{1}{1+}\ \frac{1}{13+}\ \cdots$$

...and if you are not tempted by the 13, the next partial quotient but one is 82225, giving the conjecture

$$\frac{\pi^{12}}{\zeta(12)} = \frac{638512875}{691} .$$

4.16 The best possible A_n is $\sqrt{n^2 + 4}$, the proof being essentially the same as for Hurwitz' Theorem.

4.17 The approximate location of the roots should be obvious from the first formula for $f(z)$. Applying the procedure of section 4.7 (with computer assistance if desired) gives the middle root as

$$1 + \cfrac{1}{18+} \; \cfrac{1}{5+} \; \cfrac{1}{2+} \; \cfrac{1}{5+} \; \cfrac{1}{103+} \; \cfrac{1}{1+} \; \cdots .$$

4.18 From problem 3.5, the minimal polynomial is $f(z) = z^4 - 10z^2 + 5$, and we obtain the continued fraction

$$\cfrac{1}{1+} \; \cfrac{1}{2+} \; \cfrac{1}{1+} \; \cfrac{1}{1+} \; \cfrac{1}{1+} \; \cfrac{1}{10+} \; \cfrac{1}{1+} \; \cfrac{1}{2+} \; \cfrac{1}{7+} \; \cfrac{1}{22+} \; \cfrac{1}{1+} \; \cfrac{1}{51+} \; \cdots$$

for α. Hence the required inequality is satisfied by

$$\frac{p}{q} = \frac{p_{11}}{q_{11}} = \frac{46041}{63370} .$$

4.19 Very similar to Theorem 4.22. Begin with $\coth \dfrac{1}{2p}$ instead of $\coth \dfrac{1}{2}$.

4.20 Any p/q is either a convergent to α or not: consider the two cases separately. If p/q is not a convergent, use Theorem 4.13 to find a constant c_1 such that

$$\left| \alpha - \frac{p}{q} \right| > \frac{c_1}{q^2 \log q} ,$$

provided that $q > 1$.

If p/q is a convergent p_k/q_k to α, first prove that $q_k \geq 2^{k/2}$. Dispose of the one possible exception to this inequality by recalling that $q \neq 1$. Then take the equality in equation (4.5); estimate α_{k+1} in terms of k, and hence in terms of $\log q_k$, to show that

$$\left| \alpha - \frac{p_k}{q_k} \right| > \frac{c_2}{q_k^2 \log q_k}$$

for some c_2. Then the required result holds with $c = \min(c_1, c_2)$. If α is approximable to order $s > 2$, then q^{s-2} is less than a constant times $\log q$ for arbitrarily large q, which is impossible. All of this applies to e since the partial quotients of e satisfy $a_k \leq k$ whenever $k \geq 1$.

4.21 By manipulating series we can write the modified Bessel function in terms of the function $f(c; z)$ from section 4.8,

$$I_\nu(x) = \frac{1}{\Gamma(\nu + 1)} \left(\frac{x}{2} \right)^\nu f\left(\nu + 1; \frac{x^2}{4} \right) .$$

If we now take appropriate values for ν and x, then Theorem 4.19 yields

$$1 + \cfrac{1}{2+} \; \cfrac{1}{3+} \; \cfrac{1}{4+} \; \cfrac{1}{5+} \; \cdots = \frac{I_0(2)}{I_1(2)} .$$

4.22 To evaluate I_k, write one of the factors x in the integrand as $(x-1)+1$ and split into two integrals. For K_k, integrate by parts; for J_k, use both these techniques. With careful work, this will give the recurrence relations

$$I_k = J_k + K_{k-1}, \quad J_k = -(2k+1)K_{k-1} + J_{k-1}, \quad K_k = -I_k - J_k .$$

If $k \geq 0$ we have $a_{3k+2} = 2k+2$ and $a_{3k+3} = a_{3k+4} = 1$; this leads to

$$r_{3k+2} = (2k+2)r_{3k+1} + r_{3k}$$
$$r_{3k+3} = r_{3k+2} + r_{3k+1}$$
$$r_{3k+4} = r_{3k+3} + r_{3k+2} .$$

Use all this information to prove the required formulae simultaneously by induction. A straightforward estimate shows that the integrals all tend to zero, and it follows that

$$\alpha = \lim_{m \to \infty} \frac{p_m}{q_m} = e .$$

A comprehensive exposition of the origin of these integral formulae in the work of Hermite is given by Cohn [20].

4.23 (a,b) We have $de = 1 + k(p-1)(q-1)$ for some integer $k \geq 1$; show that

$$\left| \frac{e}{n} - \frac{k}{d} \right| < \frac{k}{d} \frac{p+q}{pq} < \frac{1}{2d^2} ,$$

and use Theorem 4.13. This means that we are seeking a convergent with $q_k < \frac{1}{3}n^{1/4}$. Since the partial denominators of a continued fraction increase more or less exponentially, $q_k \approx b^k$, the maximum number of attempts needed will be something like

$$k = \frac{1}{4} \frac{\log n}{\log b} .$$

Even if n is about 10^{200}, we have only 50 or so possibilities to check – not much work with suitable software. (Of course the exact details depend on the value of b; but we only intend to give a rough argument.)

(c) We calculate the continued fraction

$$\frac{e}{n} = \frac{1}{49+} \; \frac{1}{7+} \; \frac{1}{4+} \; \frac{1}{234+} \; \cdots ,$$

noting that the partial quotient of 234 means that the q_k are increasing very rapidly and we should have a short search. The corresponding convergents are

$$\frac{p}{q} = \frac{1}{49}, \; \frac{7}{344}, \; \frac{29}{1425}, \; \frac{6793}{333794} ;$$

we can stop here as the fourth denominator is already too large. Now we should have $c^{de} \equiv c \pmod{n}$ for every c; taking $c = 2$ for simplicity, we calculate

$$2^{49e} \equiv 261918826598014\,, \quad 2^{344e} \equiv 194153471635549\,, \quad 2^{1425e} \equiv 2\,.$$

Thus only $d = 1425$ remains as a possibility; we can now find p and q to confirm that this possibility is in fact correct.

This method of breaking a carelessly set up RSA code is known as *Wiener's attack*; more information can be found in [69].

4.24 Let α have partial quotients a_k and complete quotients α_k. At the kth step we remove a_k squares, leaving a rectangle of size

$$\frac{1}{\alpha_1 \alpha_2 \cdots \alpha_{k+1}} \quad \text{by} \quad \frac{1}{\alpha_1 \alpha_2 \cdots \alpha_k}$$

(which is similar to a rectangle of size $1 \times \alpha_{k+1}$). Prove this by induction.

4.25 Since we are told that Holmes "glanc[ed] at his watch", it appears reasonable to assume that he observed the passage of (say) p telegraph posts in (say) q seconds. It seems improbable that he could achieve any greater accuracy than a single post, or a single second, so we assume that p and q are both integers. Converting a speed of p posts in q seconds to miles per hour and noting that Holmes gives the speed to the nearest half mile per hour (or better), we need

$$\left| \frac{1350p}{11q} - 53\frac{1}{2} \right| < \frac{1}{4}\,, \quad \text{that is,} \quad \left| \frac{p}{q} - \frac{1177}{2700} \right| < \frac{49}{360}\,;$$

so we look at the convergents of $1177/2700$. The first sufficiently accurate convergent is $p/q = 7/16$, the next is $p/q = 17/39$. The most plausible conclusion would seem to be that Holmes observed 7 telegraph posts in 16 seconds, and then calculated $(7/16) \times (1350/11)$ in his head. Elementary!

4.26 The figure of 29.41% suggests an extensive study involving maybe 10000 simulations. An investigation employing continued fractions might lead to the view that the actual number of tests performed was considerably smaller than this, and that the results of the simulation cannot, therefore, be considered very reliable.

CHAPTER 5

5.1 If the sides of the rectangle are $2x$ and $2y$ we get

$$\frac{4\alpha}{\pi} = t\,, \quad \frac{4\beta^2}{\pi^2} = \frac{t}{t+2} \quad \text{where} \quad t = \frac{x}{y} + \frac{y}{x}\,,$$

and this gives a quadratic for π.

5.2 Answer: $e_1^4 - 4e_1^2 e_2 + 8e_1 e_3 - 16e_4$.

5.3 Write $f(z) = \prod_{k=1}^{n}(z + a_k)$; then each side of the equation is $f'(1)/f(1)$.

5.4 Let $\alpha_1, \ldots, \alpha_n$ be complex numbers and write $\beta_k = e_k(\alpha_1, \ldots, \alpha_n)$, where e_k is the kth elementary symmetric polynomial in n variables. If at least one of $\alpha_1, \ldots, \alpha_n$ is transcendental, then at least one of $\beta_1, \ldots, \beta_n$ is transcendental.

5.6 (a) Writing $\zeta = e^{2\pi i/3}$, the conjugates are

$$\alpha_1 = \sqrt{2} + \sqrt[3]{3}, \qquad \alpha_2 = -\sqrt{2} + \sqrt[3]{3},$$
$$\alpha_3 = \sqrt{2} + \sqrt[3]{3}\,\zeta, \qquad \alpha_4 = -\sqrt{2} + \sqrt[3]{3}\,\zeta,$$
$$\alpha_5 = \sqrt{2} + \sqrt[3]{3}\,\zeta^2, \qquad \alpha_6 = -\sqrt{2} + \sqrt[3]{3}\,\zeta^2.$$

(b) The values $\alpha_1 + \alpha_2$ and so on consist of a complete set of conjugates

$$\{\pm 2\sqrt{2} - \sqrt[3]{3}\,\zeta^k \mid k = 0, 1, 2\}$$

with 6 elements; a complete set of conjugates $\{2\sqrt[3]{3}, 2\sqrt[3]{3}\,\zeta, 2\sqrt[3]{3}\,\zeta^2\}$; and the set $\{-\sqrt[3]{3}, -\sqrt[3]{3}\,\zeta, -\sqrt[3]{3}\,\zeta^2\}$ which occurs twice.

(c) Here there is a complete set of conjugates $\{\sqrt{2} + \sqrt[3]{3} + 2\sqrt[3]{3}\,\zeta, \ldots\}$ with 12 elements; and $\{3\sqrt{2}, -3\sqrt{2}\}$; and $\{\sqrt{2}, -\sqrt{2}\}$ three times.

5.7 If α is algebraic with denominator d and e^α is algebraic, consider $e^{d\alpha}$.

5.8 Using the fact that $\alpha_1 + \alpha_2 + \alpha_3 = 0$, we have

$$p(e^{\alpha_1})p(e^{\alpha_2})p(e^{\alpha_3}) = -(e^{\alpha_1} + e^{\alpha_2} + e^{\alpha_3}) + (e^{-\alpha_1} + e^{-\alpha_2} + e^{-\alpha_3}),$$

and none of these terms is an integer.

5.9 To show that (1) implies (2) rewrite the equation $\beta_1 e^{\alpha_1} + \beta_2 e^{\alpha_2} = 0$ as

$$e^{\alpha_1 - \alpha_2} = -\frac{\beta_2}{\beta_1},$$

assuming that $\beta_1 \neq 0$. For the converse, consider $1(e^\alpha) + (-e^\alpha)e^0 = 0$.

5.10 If α is algebraic and $\tan \alpha = \beta$, then $i\alpha$ is algebraic and

$$(e^{i\alpha})^2 = \frac{1 + i\beta}{1 - i\beta}.$$

Comment. There is no requirement in this question that α be real.

5.11 Evaluating the integral and the sum leads to $c^2 + (\log c - 2)c + 1 = 0$, and upon substituting $c = e^{-\lambda}$ this becomes $e^{2\lambda} - (\lambda + 2)e^{\lambda} + 1 = 0$. Writing the equation as $\lambda = -2 + 2\cosh\lambda$ makes it easy to see that there is a unique positive solution for λ. If λ is algebraic, then e^{λ} is a root of a quadratic equation with algebraic coefficients and is therefore also algebraic: this is impossible.

5.12 First prove α irrational, then use the Gelfond–Schneider Theorem.

5.13 Apply the Gelfond–Schneider Theorem to α^i and show that at least one of $\cos(\log\alpha)$ and $\sin(\log\alpha)$ is transcendental. But an exceedingly well known relation between these numbers guarantees that it is impossible for only one of them to be transcendental. **Comment.** For readers who are comfortable with complex logarithms, this result is easily extended to any complex algebraic α except for 0 and 1.

5.14 Statement (1) is equivalent to the following: if $0 < \alpha < \frac{\pi}{2}$ and $\alpha/(\pi - 2\alpha)$ is an algebraic irrational, then $2\cos\alpha$ is transcendental. Prove that (2) implies this statement.

Conversely, let β be an algebraic irrational and suppose without loss of generality that $0 < \beta < \frac{1}{2}$; consider an isosceles triangle with base angle $\alpha = \pi\beta$, and deduce (2) from the above restatement of (1).

Statement (2) follows from the Gelfond–Schneider Theorem by taking a suitable value of α.

CHAPTER 6

6.1 If d_k is the denominator of β_k, then $(d_1\beta_1)(d_2\beta_2)\cdots(d_m\beta_m)$ is an algebraic integer, and this gives an estimate for $\mathrm{den}(\beta_1\beta_2\cdots\beta_m)$; furthermore, every conjugate of $\beta_1\beta_2\cdots\beta_m$ is a product of conjugates of $\beta_1, \beta_2, \ldots, \beta_m$. For an example where equality does not hold, take $\beta_1 = \sqrt{2}$ and let β_2 be an algebraic number, not an algebraic integer, which is closely related to it.

To show that $\|\beta_1 + \beta_2\| \leq \|\beta_1\| + \|\beta_2\|$ does not always hold, we need an example where a common denominator for β_1, β_2 is greater than their individual denominators. Try thinking about some simple quadratic irrationals.

6.2 The result is obvious when $|\alpha| \geq 1$. If $|\alpha| < 1$, note that

$$a_0 = -a_1\alpha - a_2\alpha^2 - \cdots - a_n\alpha^n , \qquad (8.1)$$

so $1 \leq H(f)(|\alpha| + |\alpha|^2 + \cdots + |\alpha|^n)$, and it is easy to estimate the sum in brackets. To solve the second part of the question, turn the above inequality into an equality by choosing an α for which all the terms on the right–hand side of (8.1) are real and of the same sign.

Consider the behaviour of $|p(z)/z^k|$ in the limit as $|z|$ tends to infinity.

If e^z is not transcendental, then there is an algebraic relation which
be written

$$p_m(z)e^{mz} = p_{m-1}(z)e^{(m-1)z} + \cdots + p_0(z)$$

which holds for all z. Obtain a contradiction by showing that when
s large, the left–hand side is greater than the right–hand side in
olute value.

function

$$f(z) = \sum_{k=0}^{\infty} a_k z^k$$

sfies the functional equation $f(z) = (1+z)f(z^{10})$, and $f(z)$ can
proved transcendental using the lemmas on page 173; therefore the
litions of Theorem 6.6 apply, and α is transcendental.

the ideas of section 6.4 by considering the power series $g(z)$ with
ficients

$$b_k = \begin{cases} 1 & \text{if } k \in S_r \\ -1 & \text{if } k \notin Sr \end{cases}$$

showing that g satisfies the functional equation

$$g(z) = -(1 + z + z^2 + \cdots + z^{r-1})g(z^r) \ .$$

power series coefficients a_k satisfy

$$a_1 = 1 \quad \text{and} \quad a_{2k} = a_{2k+1} = 2a_k \text{ for } k \geq 1 \ .$$

idard series convergence tests show that f is analytic for $|z| < 1$.
ive the functional equation

$$f(z) = (2 + 2z)f(z^2) + z \ ,$$

conclude that the conditions of Mahler's transcendence theorem are
sfied. It remains to prove that f is a transcendental function. First,
w that f is not a rational function (noting that the first lemma in
endix 2.4 is not applicable). Suppose that $f(z) = p(z)/q(z)$, where
nd q are polynomials of degrees m and n respectively. Write the
ctional equation in terms of p and q, and use it to show that $m = n$.
comparing coefficients of suitable powers of z, it follows that either
or q_m is zero, which is not true. Thus f is not a rational function,
the second lemma in appendix 2.4 shows that f is transcendental

6.8 Show that the power series coefficients a_k satisfy

$$a_{3k-1} = a_{3k} = a_{3k+1} = a_k$$

for $k \geq 2$, and hence obtain the functional equation

$$f(z) = \frac{1 + z + z^2}{z} f(z^3) + z^2 - z^3 - z^4 .$$

This does not satisfy the conditions of our "functional equations and transcendence" theorem; however, considering a closely related function $g(z)$ will give a functional equation without that annoying z in the denominator.

6.9 Clearly Theorem 6.6 does not apply in this case, but we can follow very much the same method. Let $f(z)$ and ζ be as stated, and suppose that $f(\zeta)$ and $f(\zeta^2)$ are both algebraic. By assumption there is a Taylor series

$$f(z) = \sum_{k=0}^{\infty} a_k z^k \quad \text{for} \quad |z| < 1 ;$$

substituting this into the functional equation and extracting the coefficient of z^m on both sides gives

$$a_m + a_{m-1} + a_1 a_{m-2} + a_1 a_{m-3} + \cdots = \begin{cases} a_{m/4} & \text{if } 4 \mid m \\ 0 & \text{otherwise,} \end{cases}$$

and it follows by induction that all the coefficients a_k are integers. By iterating the functional equation we obtain

$$f(z^{2^t}) = \Phi_t(z) f(z)^{\phi_{t-1}} f(z^2)^{\phi_t}$$

for $t \geq 0$, where the ϕ_k are the Fibonacci numbers and

$$\Phi_t(z) = \prod_{j=1}^{t-1} \left(1 + z^{2^{t-1-j}}\right)^{\phi_j}$$

is a polynomial with integer coefficients. Show that $\Phi_t(z)$ has degree less than a constant times 2^t. Construct an auxiliary function $E(z)$ exactly as we have done throughout Chapter 6, and estimate $|E(\zeta^{2^t})|$. Show that $E(\zeta^{2^t})$ is an algebraic number whose degree is bounded independently of s and t. Considering $E(\zeta^{2^t})$ as a polynomial in three variables evaluated at $z_1 = \zeta$, $z_2 = f(\zeta)$, $z_3 = f(\zeta^2)$, estimate its algebraic size by means of Corollary 6.3. Finally, show how to choose s and t so that the estimates we have made contradict the fundamental inequality for algebraic numbers.

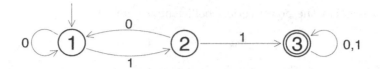

Figure 6.1 A deterministic finite automaton for exercise 6.10.

6.10 A suitable DFA is shown in figure 6.1. Using the method of section 6.5, observe that the characteristic function of S is $f(z) = f_3(z)$. If we also write $g(z) = f_2(z)$, then we can show that

$$f(z) = (1+z)f(z^2) + zg(z^2) , \quad g(z) = -zf(z^2) - zg(z^2) + \frac{z}{1-z^2} .$$

CHAPTER 7

7.1 It is convenient to also define $a_0 = 0$ and $b_0 = 1$; then we have

$$\begin{pmatrix} p_k & p_{k-1} \\ q_k & q_{k-1} \end{pmatrix} = \begin{pmatrix} a_0 & 1 \\ b_0 & 0 \end{pmatrix} \begin{pmatrix} a_1 & 1 \\ b_1 & 0 \end{pmatrix} \cdots \begin{pmatrix} a_k & 1 \\ b_k & 0 \end{pmatrix} .$$

7.2 Induction. The smart way to prove the result for q_{k+1} is to note that $q_k + q_{k-1} = (k+1)!$.

7.3 The nth convergent to the continued fraction satisfies

$$\frac{p_n}{q_n} = \sum_{k=1}^{n} \left(\frac{p_k}{q_k} - \frac{p_{k-1}}{q_{k-1}} \right) .$$

Now find a formula for q_k, use (7.5) and let $n \to \infty$.

7.4 Use the ideas of the previous problem. You should find that $q_k = k!$ for the first continued fraction, and $q_k = (2k-1)(2k-3)\cdots 1$ for the second. From the continued fraction for $\tan^{-1} x$ we obtain

$$\pi = \frac{4}{1+} \frac{1}{2+} \frac{9}{2+} \frac{25}{2+} \cdots .$$

7.5 (a) Use the fact that $q_k > q_{k-1}$ to show that $q_k > 2q_{k-1}$, then use this to show that $q_k > (2 + \frac{1}{2}b_k)q_{k-1}$. This gives rather more than was asked for.

(b) It is not hard to show that the inequality *is* true for $k = 2$, so look for an example with $k = 3$.

7.6 Let $r^2 = s/t$. Informally, we find the continued fraction

$$r \tan r = \cfrac{s}{t -} \cfrac{st}{3t -} \cfrac{st}{5t -} \cfrac{st}{7t -} \cdots .$$

So, following the procedure of section 7.2.1, define suitable a_k and b_k and deduce that p_k/q_k has an irrational limit as $k \to \infty$. To prove rigorously that the limit is $r \tan r$, show by induction that

$$rq_k \sin r - p_k \cos r = \sum_{m=0}^{\infty} (-1)^m \frac{2^k (m+k)!}{m! (2m + 2k + 1)!} \frac{s^{m+k+1}}{t^{m+1}}$$

for $k \geq 0$; the rest of the proof is very similar to that for $\tan r$.

7.7 Use induction to show that the convergents to the continued fraction are given by

$$p_k = \frac{(k+1)! \, 2^{k+1}}{[(2k-1)(2k-5) \cdots]^2} \quad \text{and} \quad q_k = \frac{[(2k+1)(2k-3) \cdots]^2}{k! \, 2^k}$$

for $k \geq 0$. Some careful algebra gives

$$\frac{p_k}{q_k} = \frac{2}{1} \times \frac{2}{3} \times \frac{4}{3} \times \frac{4}{5} \cdots \frac{2k}{2k-1} \frac{2k}{2k+1} \frac{2k+2}{2k+1} ,$$

which in the limit as $k \to \infty$ is Wallis' product for $\pi/2$.

This problem appeared in the American Mathematical Monthly in 2004; the above hints are adapted from the solution in [53].

Bibliography

[1] Boris Adamczewski and Yann Bugeaud. On the complexity of algebraic numbers. I: Expansions in integer bases. *Ann. Math. (2)*, 165(2):547–565, 2007.

[2] Martin Aigner and Günter M. Ziegler. *Proofs from THE BOOK*. Springer, 6th edition, 2018.

[3] David Angell. Arithmetic and music in twelve easy steps. *Parabola*, 26(1), 1990. Available online at `www.parabola.unsw.edu.au/1990-1999/volume-26-1990/issue-1/article/arithmetic-and-music-twelve-easy-steps`, accessed 12 May 2021.

[4] David Angell. Ordering Complex Numbers... Not. *Parabola*, 43(2), 2007. Available online at `www.parabola.unsw.edu.au/2000-2009/volume-43-2007/issue-2/article/ordering-complex-numbers-not`, accessed 29 March 2021.

[5] David Angell. Discovering Mathematics: a Case Study. *Parabola*, 45(1), 2009. Available online at `www.parabola.unsw.edu.au/2000-2009/volume-45-2009/issue-1/article/discovering-mathematics-case-study`, accessed 12 April 2021.

[6] David Angell. A family of continued fractions. *Journal of Number Theory*, 130(4):904–911, 2010.

[7] David Angell and Michael D. Hirschhorn. A Remarkable Continued Fraction. *Bulletin of the Australian Mathematical Society*, 72:45–52, 2005.

[8] Sheldon Axler. *Linear Algebra Done Right*. Undergraduate Texts in Mathematics. Springer International Publishing, third edition, 2015.

[9] Alan Baker. *Transcendental Number Theory*. Cambridge University Press, 1975.

[10] Alan Baker. *A Concise Introduction to the Theory of Numbers*. Cambridge University Press, 1984.

[11] Petr Beckmann. *A History of π*. Golem Press, second edition, 1971.

[12] Dave Benson. *Music: A Mathematical Offering*. Cambridge University Press, 2006.

[13] N.M. Beskin. *Fascinating Fractions*. Little Mathematics Library. Mir Publishers, Moscow, 1986.

[14] F. Beukers. A Note on the Irrationality of $\zeta(2)$ and $\zeta(3)$. *Bulletin of the London Mathematical Society*, 11(3):268–272, 1979.

[15] Erret Bishop. *Foundations of Constructive Analysis*. McGraw–Hill, first edition, 1967.

[16] Enrico Bombieri. Continued fractions and the Markoff tree. *Expositiones Mathematicae*, 25(3):187–213, 2007.

[17] J.W.S Cassels. *An Introduction to Diophantine Approximation*. Cambridge University Press, first edition, 1957.

[18] George Chrystal. *Algebra, an Elementary Text-Book, Part II*. Adam and Charles Black, second edition, 1900. Available online at `onlinebooks.library.upenn.edu/webbin/book/lookupid?key=olbp36404`, accessed 29 March 2021.

[19] Ruel V. Churchill and James Ward Brown. *Complex Variables and Applications*. McGraw–Hill, eighth edition, 2009.

[20] Henry Cohn. A short proof of the simple continued fraction expansion of e. *American Mathematical Monthly*, 113:57–62, 2006.

[21] Rosanna Cretney. The origins of Euler's early work on continued fractions. *Historia Mathematica*, 41(2):139–156, 2014.

[22] D. Desbrow. On the Irrationality of π^2. *The American Mathematical Monthly*, 97(10):903–906, 1990.

[23] Norman Do. Puzzle corner 3. *Gaz. Austral. Math. Soc.*, 34(3):143–148, 2007.

[24] Euclid. Elements, Book X, translation and commentary by David E. Joyce. `mathcs.clarku.edu/~djoyce/java/elements/bookX/bookX.html`, 1996, accessed 14 May 2021.

[25] L. Euler. De fractionibus continuis dissertatio. *Comm. Acad. Sci. Petropol.*, 9:98–137, 1744. Available online at `scholarlycommons.pacific.edu/euler-works/71/`, accessed 29 April 2021.

[26] James Franklin and Albert Daoud. *Proof in Mathematics: An Introduction*. Kew Books, Sydney, 2010.

[27] Jacques Hadamard. Sur la distribution des zéros de la fonction $\zeta(s)$ et ses conséquences arithmétiques. *Bull. de la Soc. math. de France*, 24:199–220, 1896.

[28] G.H. Hardy and E.M. Wright. *An Introduction to the Theory of Numbers.* Oxford, fifth edition, 1979.

[29] G.H. Hardy, E.M. Wright *et al. An Introduction to the Theory of Numbers.* OUP Oxford, sixth edition, 2008.

[30] C. Hermite. Extrait d'une lettre de Mr. Ch. Hermite à Mr. Borchardt. *J. de Crelle*, 76:342–344, 1873.

[31] C. Hermite. Sur la fonction exponentielle. *Compt. Rend. Acad. Sci. Paris*, 77:18–24, 74–79, 285–293, 1873.

[32] David Hilbert. Mathematische Probleme. *Nachrichten von der Gesellschaft der Wissenschaften zu Göttingen*, pages 254–297, 1900.

[33] David Hilbert. Mathematical Problems. `mathcs.clarku.edu/~djoyce/hilbert/problems.html`, 1996. A translation of [32], accessed 22 June 2021.

[34] James Jeans. *Science and Music.* Cambridge Library Collection – Mathematics. Cambridge University Press, 2009. First published 1937.

[35] A. Ya. Khinchin. *Continued Fractions.* State Publishing House of Physical–Mathematical Literature, third edition, 1951. English translation 1964.

[36] Stephan Körner. *The Philosophy of Mathematics: An Introductory Essay.* Dover, 1986.

[37] Johann Heinrich Lambert. Mémoire sur quelques propriétés remarquables des quantités transcendentes circulaires et logarithmiques [Memoir on some remarkable properties of transcendental circular and logarithmic quantities]. *Mémoires de l'Académie Royale des Sciences de Berlin*, 17:265–322, 1768. Written in 1761 but published only in 1768. Available online at `www.kuttaka.org/ℲHL/L1768b.html`, accessed 29 April 2021.

[38] Ferdinand Lindemann. Ueber die Zahl π [On the number π]. *Mathematische Annalen*, 20:213–225, 1882. Available online at `www.digizeitschriften.de/dms/img/?PID=GDZPPN002246910`, accessed 29 April 2021.

[39] Joseph Liouville. Sur des classes trés étendues de quantités dont la valeur n'est ni algébrique, ni même réductible à des irrationnelles algébriques. *J. Math. Pures Appl.*, 16:133–142, 1851.

[40] Joseph Lipman. *Transcendental Numbers.* Number 7 in Queen's Papers in Pure and Applied Mathematics. Queen's University, Kingston, Ontario, 1966.

[41] K. Mahler. Arithmetische Eigenschaften einer Klasse von Dezimal-brüchen. *Proc. Akad. Wet. Amsterdam*, 40:421–428, 1937.

[42] K. Mahler. On the approximation of π. *Nederl. Akad. Wet., Proc., Ser. A*, 56:30–42, 1953.

[43] K. Mahler. Fifty years as a mathematician. *Journal of Number Theory*, 14(2):121–155, 1982.

[44] Ivan Niven. A simple proof that π is irrational. *Bulletin of the American Mathematical Society*, 53(6):509, 1947.

[45] Ivan Niven and Herbert S. Zuckerman. *An introduction to the theory of numbers.* Wiley, New York, third edition, 1972.

[46] Ivan Niven, Herbert S. Zuckerman and Hugh L. Montgomery. *An Introduction to the Theory of Numbers.* Wiley, fifth edition, 1991.

[47] C. D. Olds. The Simple Continued Fraction Expansion of e. *The American Mathematical Monthly*, 77(9):968–974, 1970.

[48] Adam C. Orr. More Mathematical Verse. *The Literary Digest*, 32(3):83–84, 1906.

[49] Ignacio Palacios–Huerta. Tournaments, fairness and the Prouhet–Thue–Morse sequence. *Economic Inquiry*, 50(3):848–849, 2012.

[50] Stephen G. Penrice. Problem 1566. *Mathematics Magazine*, 72(1):64, 1999.

[51] G. Pólya and G. Szegő. *Problems and Theorems in Analysis,* volume 2. Springer, 1972.

[52] George Polya. *How to Solve It: A New Aspect of Mathematical Method.* Princeton University Press, 1945.

[53] Leroy Quet and Peter Ørno. A Continued Fraction Related to π. *Amer. Math. Monthly*, 113(6):572–573, 2006.

[54] Denis Roegel. Lambert's proof of the irrationality of Pi: Context and translation. Includes a translation of [37]. Research report, LORIA, 2020.

[55] Barkley Rosser. Explicit Bounds for Some Functions of Prime Numbers. *American Journal of Mathematics*, 63:211, 1941.

[56] K.F. Roth. Rational approximations to algebraic numbers. *Mathematika*, 2:1–20, 168, 1955.

[57] V. Kh. Salikhov. O mere irratsional'nosti chisla π. *Usp. Mat. Nauk.*, 63(3):163–164, 2008.

[58] V. Kh. Salikhov. On the measure of irrationality of the number π. *Mathematical Notes*, 88:563–573, 10 2010. A translation of [57].

[59] Carl Ludwig Siegel. Approximation algebraischer Zahlen. *Math. Zeitschrift*, 10(3):173–213, 1921.

[60] Daniel Solow. *How to Read and Do Proofs*. Wiley, sixth edition, 2014.

[61] Ian Stewart. *Galois Theory*. Chapman and Hall, 1973.

[62] Ian Stewart and David Tall. *Algebraic Number Theory*. Chapman and Hall, London, first edition, 1979.

[63] Axel Thue. *Om en generel i store hele tal uløsbar ligning*. Skrifter udg. af Videnskabs–selskabet i Christiania. I, Math.–naturv. klasse. Jacob Dybwad, 1908.

[64] Axel Thue. Über Annäherungswerte algebraischer Zahlen. *Journal für die reine und angewandte Mathematik*, 135:284–305, 1909.

[65] de la Vallée Poussin, Charles–Jean. Recherches analytiques sur la théorie des nombres premiers. *Ann. de la Soc. scientifique de Bruxelles*, 20:183–256, 281–397, 1896.

[66] Alfred van der Poorten and R. Apéry. A proof that Euler missed... – An informal report. *Mathematical Intelligencer*, 1(4):195–203, 1979.

[67] K. Weierstrass. Zu Lindemann's Abhandlung: "Über die Ludolph'sche Zahl". *Sitzungsberichte der Königlich Preussischen Akademie der Wissenschaften zu Berlin*, 5:1067–1085, 1885.

[68] Jacob Westlund. On the Irreducibility of Certain Polynomials. *The American Mathematical Monthly*, 16(4):66–67, 1909.

[69] M.J. Wiener. Cryptanalysis of short RSA secret exponents. *IEEE Transactions on Information Theory*, 36(3):553–558, 1990.

[70] M. Wyman and B. Wyman. An essay on continued fractions. *Math. Systems Theory*, 18:295–328, 1985. A translation of [25]. Available online at kb.osu.edu/handle/1811/32133, accessed 29 April 2021.

[71] Doron Zeilberger and Wadim Zudilin. The irrationality measure of π is at most 7.103205334137.... *Moscow Journal of Combinatorics and Number Theory*, 9(4):407–419, 2020.

Index

Printed in the United States
by Baker & Taylor Publisher Services